AF798838

Palgrave Studies in Financial Services Technology

Series Editor
Bernardo Nicoletti
Rome, Roma, Italy

The Palgrave Studies in Financial Services Technology series features original research from leading and emerging scholars on contemporary issues and developments in financial services technology. Falling into 4 broad categories: channels, payments, credit, and governance; topics covered include payments, mobile payments, trading and foreign transactions, big data, risk, compliance, and business intelligence to support consumer and commercial financial services. Covering all topics within the life cycle of financial services, from channels to risk management, from security to advanced applications, from information systems to automation, the series also covers the full range of sectors: retail banking, private banking, corporate banking, custody and brokerage, wholesale banking, and insurance companies. Titles within the series will be of value to both academics and those working in the management of financial services.

More information about this series at
http://www.palgrave.com/gp/series/14627

Peter Scholz
Editor

Robo-Advisory

Investing in the Digital Age

palgrave
macmillan

Editor
Peter Scholz
Hamburg School of Business
Administration
Hamburg, Germany

ISSN 2662-5083 ISSN 2662-5091 (electronic)
Palgrave Studies in Financial Services Technology
ISBN 978-3-030-40817-6 ISBN 978-3-030-40818-3 (eBook)
https://doi.org/10.1007/978-3-030-40818-3

© The Editor(s) (if applicable) and The Author(s), under exclusive license to Springer Nature Switzerland AG 2021
This work is subject to copyright. All rights are solely and exclusively licensed by the Publisher, whether the whole or part of the material is concerned, specifically the rights of translation, reprinting, reuse of illustrations, recitation, broadcasting, reproduction on microfilms or in any other physical way, and transmission or information storage and retrieval, electronic adaptation, computer software, or by similar or dissimilar methodology now known or hereafter developed.
The use of general descriptive names, registered names, trademarks, service marks, etc. in this publication does not imply, even in the absence of a specific statement, that such names are exempt from the relevant protective laws and regulations and therefore free for general use.
The publisher, the authors, and the editors are safe to assume that the advice and information in this book are believed to be true and accurate at the date of publication. Neither the publisher nor the authors or the editors give a warranty, expressed or implied, with respect to the material contained herein or for any errors or omissions that may have been made. The publisher remains neutral with regard to jurisdictional claims in published maps and institutional affiliations.

This Palgrave Macmillan imprint is published by the registered company Springer Nature Switzerland AG.
The registered company address is: Gewerbestrasse 11, 6330 Cham, Switzerland

For Gwendolyn and Nola

Foreword

Man or machine?

This is one of the key questions of our time. Albeit not new in its historical context, it is once again being heavily discussed—now in the context of the "labor market". Machines make certain tasks superfluous, whilst at the same time creating new professions and thus new jobs. And yet, one question remains: Can the number of jobs created compensate for the ones lost? What will the future look like for our work and our lives? When it comes to assembly line robots, automation technology or artificial intelligence, people today tend to have a somewhat negative bias of discomfort, mistrust and perhaps even fear toward machines. And on top of it all, they are now asked to entrust their financial security to a machine.

As always, though, there are two sides to the coin: As a matter of fact, we all enthusiastically use smart machines every day. We can no longer imagine life without a laptop, smartphone, Thermomix or an automated vehicle kitted out with driver-assisting technologies to increase road safety and convenience. So why not entrust our money to smart machines?

Especially when they offer professional asset management with little effort for even small investments. "Professional increase of money for everyone"—such robo-advisor solutions may at first sound a bit like a socialistic approach attempting to penetrate the exclusive world of asset management, until then only available to people with at least six- or seven-figure investment sums. When they entered the market a few years ago though, providers announced no less than a technical and content revolution, the disruption of traditional asset management and financial advice. Investments would grow in a cost-effective, simple and automated way, globally diversified and secured 24/7, with a state-of-the-art risk management.

The "financial advisor in the machine" acts without emotions, solely following an algorithm. In theory, this should eliminate classical behavioral finance errors, which private and professional investors alike are often subjected to. Ideally, ruling out the principle of the herd instinct—buying too late and selling too early, panicking, and above all falling at risk of emotionally charged decision making. "Simply ingenious", as many a financial blog dubbed this new option. Sounding too much like an out-and-out benefit for mankind? As it goes, there is of course a concise business strategy behind it: using modern technology to gain market shares from traditional asset managers.

The robo-advisor technology in principal consists of systems professional asset managers have been using for more than 20 years to recognize market movements early on and thus act on them in a timely manner. At its core, it is about gaining a competitive advantage of information. Financial market professionals have far more options than private investors when it comes to observing and analyzing markets and the occasionally occurring, irrational behavior of their participants. Computer programs play an essential role in this. They recognize patterns and anticipate developments, thus making it possible to act at once, instead of having to react as victims of market movements.

The underlying principle is the same as that which was practiced centuries ago on the London Stock Exchange. Back in those days, there was always an uncertainty as to whether cargo vessels would make their way to England. If, for example, they had passed the Cape of Good Hope unscathed, observers would report this back home. The information advantage thus gained had a direct impact on share prices. Today, satellite images and data centers are used for this purpose. Then as now, the correct evaluation of data is fundamental.

This may well be the reason why even professionals can by no means always beat the market, even if such computer systems have been at their disposal for a long time. This gives rise to questions that are of course of essential importance to the entire financial sector: Can machines alone be considered a cure-all? Are purely technologically controlled investments in fact a more lucrative option? Can they reduce risks and increase opportunities? Will bank advisors and asset managers become redundant in the foreseeable future? Are robo-advisors really the revolution? And if so, why have only a handful of robots been able to establish themselves on the market so far?

This book is the first to provide comprehensive answers to these questions in a fundamental, decisive, detailed and nuanced way. It clarifies the basics, the technology and the tactics behind those clever, financial machines, gives insights into their previous track record to date and much more. Looking ahead, it provides a preview of what is and may be yet to come. As a matter of fact, so far only a relatively small percentage of the global investment community have more or less relied on robo-advisors, depending on their respective culture. It is also a fact that we are only at the beginning of development. We have all borne witness to how exponentially fast things can move forward. One such example is the evolution of smartphones—which by the way have been around for just a little longer than robo-advisors.

The business model of banks and asset managers has remained unchanged and immovable for centuries. Given the possibilities arising from new technologies, FinTechs such as robo-advisors have created a completely new situation for themselves, in which no stone seems to have been left unturned. The financial business world of today is clearly a hugely exciting place and by no means solely for professionals working in or dealing with it.

"Man or machine?"—This question really does concern everyone. We live in times of rapidly exponential and radical change. It is a truth universally acknowledged that fear is a bad advisor. Confidence, on the other hand, makes all progress possible and tangible.

Therefore, the answer to this fundamental question can only be:

"Man with machine."

Munich, Germany
March 2020

Jochen Werne

PREFACE

Digitalization—it is a megatrend that has accompanied our economy for many decades now. Although one might think that the smartphone and mobile internet have created digitalization, it roots back to Konrad Zuse, who invented the first working computer in 1941. Since then, computational power has developed at a rapid pace, like predicted by Moore's Law, and accelerated the automatization tendency of industrialization. The introduction of personal computer in our daily working routine is considered as industry 3.0 and was strongly driven by Apple and Microsoft. Only a few years later, during New Economy and internet bubble around the turn of the millennium, the internet conquered the markets and showed what potential lies within computer networks. This has been the starting point for the industry 4.0. Today, smartphones and mobile internet allow users to enjoy the benefits of global computer networks anytime and anywhere. The *FAANG* companies, Facebook, Amazon, Apple, Netflix, and Google hence are dominating the economy and have disrupted many industry sectors. Latest since the financial crisis from 2008, due to the loss in trust in the financial system, the banking sector as well cannot hide from the digital age anymore and needs to adopt to the new consumer behavior, which expects mobile access to banking services 24/7 without restrictions in location. Fintechs already have discovered this field and create innovative products and services.

Since the New Economy, I am closely watching the digital transformation of the banking sector. From the first pure online banks in the nineties until today, when Fintechs and cryptocurrencies are trying to establish at financial markets: the development is tremendous and it is a privilege to be a part of it. But the digital evolution also comes with many challenges. In my role as a professor for banking and financial markets, I have to prepare and enthuse my students for new requirements in the field, which includes the natural application of software and basic programming skills. This is crucial, not because they are expected to really code but to be able to explain to the programmers what to implement and to understand at least the core of the problems the IT experts are facing, like data integrity and the configuration of interfaces. IT projects will be an essential part of banks in future: the application of more and more software solutions needs more bankers who are able to precisely define and describe processes: for example, how a machine can provide excellent investment advice.

The digitalization of the banking sector also opens new and interesting research opportunities. Together with my MBA alumnus Michael Tertilt, we have analyzed how robo-advisors measure the risk profile of investors and how they derive a portfolio recommendation from this information. We have been surprised that they seem to ask way more questions than they really use for the process of deriving a portfolio for the investors. In sum, we think that the quality of the digital advice is not worse than the average level in the market, but we still see much potential for developments in the future. As I presented our findings at the 54th Meeting of the Eastern Finance Association in Philadelphia (USA) in April 2018, I was approached by Palgrave Macmillan and asked if I want to write a book about robo-advice. I was immediately thrilled by the idea, but I also recognized the challenge to write a book about such a complex topic since robo-advice is an interdisciplinary subject. Hence, the idea was born to act as an editor and to invite different experts in the field to contribute to the book. Since robo-advisory is a topic that was born in the business life and not in academia, the current knowledge is widely distributed; therefore, we need to build a bridge in the book from descriptions of best practice to considerations of applied sciences. This is why I divided the book in four different parts.

The first part of the book explores the current status quo of robo-advisory: in Chap. 1 we define the term robo-advice and analyze its clients and market (potential); we describe the wealth management process in which robo-advisory aims to participate; we briefly look back in the history of robo-advisory; and we analyze the past performance of robo-advisors—probably the ultimate long-term criteria for clients. Furthermore, the current status quo based on the academic literature is surveyed in Chap. 2.

The second part of the book deals more with the implementation of robo-advisory. Risk profiling is one of the crucial elements of the advisory process and a component which can really add value for clients. But as we learned from our paper (Tertilt and Scholz 2018), it is difficult to efficiently measure the investor's risk profile. Hence, we include Chap. 3, which explains how the insights from psychometrics can help to improve risk profiling of robo-advisors. As we will learn in Chap. 1, the portfolio management process of robo-advice still relies widely on investment committees and there seems generally more potential for human interference than expected. In Chap. 4, we will discuss and observe the potential implications and emotional biases that may occur based on not wholly automatized investment process. But we will also see that, especially US robo-advisors, apply the concept of goal-based investing and thus seem to have a more client-centered angle. Chapter 5 introduces the concept of rule-based investing and will show how human impact in the portfolio management can be further reduced since robo-advice is privileged to apply fully automated processes. Another important aspect is the acceptance of human clients to be advised by machines, which will be addressed in Chap. 6 Moreover, regulation is also an important part of the investment advisory business. Especially since the financial crisis, the regulation has been severely tightened and the development of robo-advisory has further created the need to adapt regulation; meanwhile regulation has become a field for true experts. Since the regulation can be very different, we include two chapters on the topic: one for our US readers (Chap. 7) and one for our European/German readers (Chap. 8). Although the European regulations can differ in its intricacies from country to country, many rules are defined on European level and have been transferred to national law.

The third part of the book contains case studies of best practice. The first contribution in this part (Chap. 9) is a whitepaper by BearingPoint and describes from the perspective of experienced consultants the most important elements which need to be considered if a robo-advisor is incorporated. The second chapter (Chap. 10) is an extensive report by the traditional bank Hauck & Aufhaeuser and explains the different challenges they faced and the solutions they found in their project to introduce a robo-advisor for their clients. Both contributions give valuable insights in the practical process of digital transformation.

The fourth part looks ahead to the near future of robo-advisory. It seems to be a natural assumption that robo-advisory is a perfect field for big data and artificial intelligence (AI). Today, however, these technologies are not widely spread amongst robo-advisors. Chapter 11 introduces the topic of AI and describes the requirements and possibilities of robo-advice in this field. Whereas the client communication has large potential for AI applications, the improvement of the portfolio management, especially in forecasting, seems to be rather limited. In this regard, Chap. 12 analyzes the marketing impact and data gathering potential of social networks for the robo-advisory business. Finally, Chap. concludes by summing up the most important factors, which seem to determine the success of a robo-advisor in the future.

We created the book for a diverse readership: from students, who want to speed up their knowledge in digital transformation and investment advisory topics, to academics, who want to gather a profound overview about robo-advice, to practitioners, who may seek intellectual support for how to implement a robo-advisor, and last but not least, to those investors, who are highly interested in the topic and want to learn more about the technology and process of robo-advice. I am grateful for working together with all contributing authors and I am confident that we created a book which is enriching for our readers.

Hamburg, Germany Peter Scholz
May 4th, 2020

Reference

Tertilt and Scholz (2018) Michael Tertilt and Peter Scholz. 2018. "To Advise, or Not to Advise—How Robo-Advisors Evaluate the Risk Preferences of Private Investors". *The Journal of Wealth Management* 21(2): 70–84.

Acknowledgements

A large project like this book on robo-advisory, where so many perspectives are brought together, can never be achieved alone. Therefore, I am grateful to all authors that contributed their expertise to create such a differentiated picture of such a complex topic. Dear Alexander D. Beck, Ana-Maria Climescu, Melanie L. Fein, Joachim Goldberg, Goetz Greve, David Grossmann, Christian Hammer, Christian von Keitz, Sinan Krueckeberg, Frederike Meyer, Monika Mueller, Yvonne Quint, Paul Resnik, Jan Rocholl, Thorsten Ruehl, Madeleine Sander, Craig Saunders, Theodor Schabicki, Soeren Schroeder, and Michael Tertilt: thank you so much for your valuable contributions! I really appreciate your effort and time! Please allow me to extend a special thank you for the smooth teamwork to my co-authors Joachim Goldberg, David Grossmann, and Michael Tertilt. Another big and special thank you goes to Theodor Schabicki from BearingPoint, who permitted to use their whitepaper as a case study in this book! I am really looking forward to further cooperation with you!

But support is also necessary behind the scenes. I would like to thank Jochen Werne in particular, since he made contacts with other authors, who then contributed. Moreover, I would like to thank you for your inspiring foreword. To know such a highly sophisticated networker is a real privilege.

Another privilege I really appreciate is the cooperation with my former Ph.D. students David Grossmann and Sinan Krueckeberg. We went together through many research projects in the past and yet teamed up again. Both of you have contributed as authors, and both of you have provided constructive and valuable feedback to my contributions. Thank you very much for your friendship and support!

Furthermore, I owe a big thank you to my student assistant Laura Andrade Pardo, who assisted me in the LaTeX transfer and proper referencing of the bib-file. My colleague Franziska Buttler provided assistance in organization and administrative work: thank you for keeping me free of many administrative ties! The book is written in LaTeX: here, another big thank you goes to my friend Lars Geiger for his support in troubleshooting if the code did not work and the error seemed to be impossible to find. He really is a lender of last resort!

I am also grateful to my publisher Palgrave Macmillan, who contacted me and gave me the opportunity to edit a whole book on such an interesting and up-to-date topic. Thank you in particular for the patience and support during this project. Although it has been a true peak in workload, I really enjoyed to deep-dive into the field of robo-advice! Support from my employer HSBA is gratefully acknowledged.

Finally, and most importantly, I am thankful to my wife Britta for her support and understanding. Above all, I am most grateful to my wonderful daughters Gwendolyn and Nola for being the light in my life.

Contents

Notes of Contributors

Alexander Beck is an experienced technology expert in the field of financial markets forecasting. During his academic career in physics and economics, he adapted methods of statistics and machine learning from the field of high energy physics to understanding and forecasting financial markets. He holds a master's degree in physics and a Ph.D. in economics. During his career as a consultant, he advised several financial services companies in the application of machine learning and artificial intelligence. He is the CTO of the quantitative asset manager Quoniam Asset Management.

Ana-Maria Climescu is Senior Associate at Hauck & Aufhäuser and is responsible for marketing. She has worked for the bank since 2015 and during this time, she has developed marketing campaigns for various business areas and accompanied projects. She is also responsible for the entire external appearance of H&A's digital wealth channel Zeedin. Since 2019, she has also worked as a lecturer for external corporate communications. Climescu holds an M.A. in Global Marketing Management from Accadis University Bad Homburg.

Melanie L. Fein is an attorney who advises financial institutions on regulatory matters. She formerly was senior counsel to the Board of Governors of the Federal Reserve System and has served on the adjunct faculty of Yale Law School. Portions of this chapter have been previously included in papers by Fein and are available at the Social Sciences Research Network (SSRN).

Joachim Goldberg has analyzed the interaction between investors and markets for more than 40 years. The banker and former currency trader was inspired by the insight that psychological factors drive financial markets. Since then, he took a deep look into behavioral finance, a discipline which analyzes the emotional decision-making process of market participants.

In 1999, together with Professor Rüdiger Nitzsch of RTHW Aachen, he published the book *Behavioral Finance – Gewinnen mit Kompetenz*, which is one of the first books on the subject of the German market. Through many appearances on radio and TV, press reports, internet columns as well as numerous events for behavior-oriented market analysis, Goldberg has established himself as a leading expert in this field in Germany and is considered one of its best-known representatives.

Goetz Greve is Professor of marketing and sales at the HSBA Hamburg School of Business Administration, Germany. In his research, he focuses on the areas of customer relationship management, sales management, and online marketing. His research findings have been published in international journals such as the *International Journal of Research in Marketing* and *Marketing Review St. Gallen*. Goetz has published four books in the field of CRM and Online Marketing. He is on the editorial boards of the *International Journal of Internet Marketing and Advertising* and the *International Journal of Marketing Studies* and serves as an ad-hoc reviewer for other journals, such as the *European Journal of Marketing* and the *Journal of Marketing Theory and Practice*. Before returning to academia, he worked for Accenture and OC&C Strategy Consultants.

David Grossmann earned his Ph.D. from the Andrássy University Budapest—University of National Excellence—in cooperation with the Claussen-Simon Graduate Centre at HSBA. He is the Head of Global Bank Management at the Otto M. Schröder Bank AG. Formerly, he was with the Hamburger Sparkasse AG, a significant European institution, where he worked as a professional in Real Estate Finance, Enterprise Risk Management, and Performance Management.

Christian Hammer worked as a savings bank manager for private and corporate clients for 10 years before moving to NFS in 2005. As a graduated bank manager and financial planner (HfB) he looked for the dynamic and innovative challenge in the banking landscape and found his home as an Executive Partner at NFS. By writing ***Basic knowledge for investment funds, securities and alternative investment funds***, he passed on his many years of experience in the capital markets to a larger audience.

Christian von Keitz is Senior Project Manager in Hauck & Aufhäuser's Inhouse Consulting department. He has more than 12 years of profound experience in project management within the banking business (specialized in IT driven projects with a private wealth, regulatory or business development related background). von Keitz joined Hauck & Aufhäuser in 2017, and within the Zeedin project team, he has been responsible for the conception of the customer self-assessment, the mechanisms for the resulting investment proposal as well as for several aspects of the onboarding process. He started his professional career with a vocational training in banking before pursuing an international management degree (B.A.).

Sinan Krueckeberg has more than 10 years of experience in conventional and alternative investments in practice, research, and teaching across the US, Europe, and Asia.

As CEO of the Krueckeberg Family Office, he is responsible for the family's investment strategy which has allowed him to gain a deep understanding of the dynamics and challenges of both public and private markets. He also leads the Private Capital practice of Etribes, a leading European digital strategy consultancy that helps the private equity industry increase portfolio company valuations through digital levers.

As executive board member of a large mid-sized family business operating in the paper trading, packaging and screen & sign industries, he has first-hand experience of the threats and opportunities that digitization creates for legacy portfolio companies.

He complements his practical experience with academic research into alternatives and his work has most recently been published in the internationally renowned CFA Institute's *Financial Analysts Journal*. Sinan is also a senior lecturer at the ***Hamburg School of Business Administration*** where he lectures on Entrepreneurial Finance in the M.Sc. Finance program.

Frederike Meyer lives in Hamburg and works in a wealth management team of a regional financial institute where she is responsible for corporate banking. She finished her bachelor's degrees in educational science and business administration. She is pursuing her master's degree in business psychology.

Monika Mueller founded FCM Finanz Coaching in 1999 and works as an Executive Coach mainly with CEOs, directors and financial sales, portfolio- and management teams of German and international companies in the field of financial services. She offers coaching in the areas of organizational and personal transformation and effectiveness, team intelligence, development of conscious leadership competencies. She also offers training and coaching in financial decision making for portfolio managers, financial advisors, and HNWI. She is an expert in risk profiling and brain-friendly decision making.

Mueller is considered as the pioneer of finance coaching and its development in Germany and other European countries. She was one of the first coaches in Germany to obtain certification with ICF (International Coach Federation). She is a Master Certified Coach since 2003. She did her coach training at several institutes in Europe starting from 1989. She was trained as systemic coach (WISPO), ontological coach (Newfield), EMDR coach (Sam Foster) and has also conducted studies in body therapy, energy work and organizational development. Her commitment to the dissemination of coaching led her to be a founding member of the German Chapter of ICF in 2000. Since then she is engaged in mentor coaching and the ICF assessment of coaches wordwide. In 2013, she started the FCM Finanz Coach® training program in Wiesbaden.

With a Diploma in Psychology from the Johannnes-Gutenberg University (Mainz, Germany), Mueller was head of a training and coaching agency of a German training institute and founded FCM Finanz Coaching 1999, a company that is focused on high performance in (financial) decision making.

Yvonne Quint is a partner in the segment Banking & Capital Markets at BearingPoint, and advises banks and investment companies regarding diverse capital markets and investment management topics. Her core areas of consulting are regulation of capital markets/capital markets compliance and optimization/digitalization of front-to-back-office processes for capital markets products. Her client base is broad and encompasses state banks, private banks, large international banks as well as investment companies. Over the last 4 years, Quint supported various clients in implementing

MiFID II and PRIIPs. SFTR is one of the crucial topics on her agenda. Before starting at BearingPoint in mid-2010, Quint had gained practical experience over several years in banks. She holds an Executive M.B.A. dual-degree from EBS Business School and Durham University Business School in the UK.

Paul Resnik is a founder of the world's first psychometric financial risk tolerance tool for financial advisers. The FinaMetrica Risk Profiler has been used in more than 20 countries for nearly 25 years and is now part of the Morningstar group.

Resnik is CEO of the Suitable Advice Institute which advances financial advice that is customized for the client and founded on their informed consent, based on their full understanding of risk. Find out more at suitableadviceinstitute.com.

Jan Rocholl is Director at Hauck & Aufhäuser and heads the Inhouse Consulting. He has more than 18 years of project and consulting experience in banking, industry and information technology. Rocholl has been with Hauck & Aufhäuser since 2011 and has managed and implemented a wide range of strategic and procedural projects. He was co-project manager for H&A's new Online Banking platform and for the development of H&A's digital wealth channel Zeedin with focus on all IT related matters. Prior to Hauck & Aufhäuser, he gained his project competence as a consultant for Accenture and in various positions within the SAP Group. Rocholl holds a master's degree in business administration from the University of Wuerzburg.

Thorsten Ruehl is partner and head of research and portfolio management at CSR Beratungsgesellschaft in Hofheim/Ts since 2014. Earlier, he was heading the quant asset allocation team at Deka Investment in Frankfurt for more than 10 years where he was responsible for the development of quant asset allocation strategies and their implementation in funds.

Ruehl studied Math and Physics at the Justus-Liebig-University in Gießen and at the University of Washington in Seattle. He completed his doctorate in applied solid state physics at the Institute of Applied Physics at the Justus Liebig University.

Ruehl has two decades of experience in quantitative portfolio management and is an expert for the implementation of the Black-Litterman-algorithm and portfolio insurance strategies. He gives lectures in mathematics and finance at the HS-RM in Wiesbaden and the HSBA in Hamburg.

Madeleine Sander is Managing Director and Chief Financial Officer at Hauck & Aufhaeuser. In this function, she is responsible for the departments of Corporate Development, Finance & Tax as well as H&A's digital wealth channel Zeedin. She joined Hauck & Aufhäuser in 2017 as Head of Corporate Development and was co-project manager for H&A's new Online Banking Platform and the development of H&A's digital wealth channel Zeedin with focus on all strategy related matters. Prior to Hauck & Aufhäuser, she worked more than 6 years for Deutsche Bank in the Corporate Development (AfK) and Inhouse Consulting department. This math and business administration graduate started her career at DekaBank as Business Manager for Corporates & Markets.

Craig Saunders is a communications and marketing consultant who specializes in telling stories of personal finance and investment. He has an extensive background as a financial journalist with Australia's national TV broadcaster and created and delivered an ongoing professional development program to an Australia-wide audience of financial advisers. Find out more at cbkcommunications.com.au.

Theodor Schabicki has worked in financial services for 24 years, of which the last 15 years have been at leading consulting engagements at BearingPoint, where he became Partner in 2016. As a former securities trader, his project activities cover capital market products and processes. With the BearingPoint competence team "Capital Markets – Securities" his focus lies on E2E processes, Trading, Clearing and Settlement Platforms, Cost Reduction Programs, Market Data Management and new technologies such as Distributed Ledgers.

This technical focus is complemented by his activities around process management. Since 2007, Schabicki is the leader of the competence team "Process Optimization & Design" with its main focus on activities in Process Strategy and optimization methodologies, Business Process Management (BPM), Process Mining, Robotics & Process Automation (RPA) and Artificial Intelligence (AI), Business Process Outsourcing (BPO) and holistic concepts like Data Driven Operations in Banking.His international experience spreads over engagements in Tokyo, London, Paris, Milan, Vienna, Geneva and Zurich.

Peter Scholz is Professor of Banking and Financial Markets at Hamburg School of Business Administration (HSBA) where he has been a faculty member since 2013. He is the academic head of the MSc Finance program and responsible for developing online teaching methods at HSBA. Scholz got his PhD at Frankfurt School of Finance and Management where he worked in the Centre for Practical Quantitative Finance. His research interest is focused on digitalization aspects of financial markets such as fintechs and cryptocurrencies. Furthermore, he is active in the fields of behavioral finance and active asset management. His work is published in academic journals such as the *Financial Analysts Journal* and the *Journal of Wealth Management*.

Moreover, Scholz has practical experience in the financial sector. He started his career with Deutsche Bank where he learned the business from the scratch: from investment advisor to equity derivatives sales trader. He has also worked in the quantitative equity division as senior portfolio manager for Deka Investments.

Soeren Schroeder is a manager at BearingPoint with 10 years of experience in Financial Services. He advises banks regarding transformation and process optimization topics and is part of the BearingPoint competence team "Capital Marketes – Securities". His core areas of expertise are digitalization of front-office processes and management of IT-Transformation projects. Over the last 6 years Schroeder supported various clients in implementing new product/service offerings and transforming their IT-Infrastructure.

Before starting at BearingPoint in 2013, Schroeder worked for a large retail bank. He graduated from ESADE Business School and is a Chartered Financial Analyst (CFA).

Michael Tertilt holds a bachelor's degree in Technology and Business and worked since 2004 in several positions at Diebold Nixdorf in the field of IT outsourcing services for financial industry. He completed an M.B.A. at the Hamburg School of Business Administration in 2016, specializing in bank management. He has been following the developments in the Fintech industry for several years and wrote his M.B.A. thesis and publications on Robo-Advisors.

Jochen Werne is a German manager with a proven track record in business & digital transformation and in the field of People's Diplomacy. He is a member of the Executive Committee of PROSEGUR Cash Services and is in charge of Business Development, Innovation, Business Process Outsourcing, Solutions, New Products and International Sales. He is a member of the Federal Ministry of Education & Research artificial intelligence initiative "Learning Systems"; member of one of the world's leading think tanks, the Royal Institute of International Affairs "Chatham House". He is a recognised keynote speaker and co-author of several books about innovation, leadership, AI, FinTech, Cyber Security and so on. He has been awarded several times for his engagement in the support of international relations. He is recognised by Focus-Magazine as one of the leading AI experts in Germany. He is the ambassador of the Peter Tamm sen. Foundation, the founder of the Global offshore Sailing Team (GOST) and co-founder of the NGO Mission4Peace, which is dedicated to historic research, the establishment of international, diplomatic relations and the encouragement of international dialogue. For his efforts, Werne has repeatedly received international recognition.

Acronyms

AI	Artificial Inteligence
AISP	Account Information Service Providers
API	Application Programming Interface
ARC	Asset Risk Consultants
AuM	Assets under Management
B2B	Business to Business
BaFin	Bundesanstalt für Finanzdienstleistungsaufsicht (Federal Agency for Financial Services Supervision)
CAC	Customer Acquisition Costs
CAGR	Compound Annual Growth Rate
CAPM	Capital Asset Pricing Model
CEO	Chief Executive Officer
CD	Corporate Design
CLV	Customer Lifetime Value
CSIA	Charles Schwab Investment Advisory
CSIT	Customer-Specific Interaction Technologies
DIY	Do It Yourself
DOL	Department of Labor
DR	Diversification Ratio
ERISA	Employee Retirement Income Security Act of 1974
ESMA	European Securities and Markets Authority
ETC	Exchange Traded Commodities
ETF	Exchange-Traded Fund
ETN	Exchange Traded Notes

FINRA	Financial Industry Regulatory Authority
FinTechs	Financial Technologies
FinVermV	Investment Brokerage Regulation
FTSE	Financial Times Stock Exchange
GewO	German Trade Regulation Act
GDPR	General Data Protection Regulation
GRNN	General Regression Neural Network
GWG(AMLA)	Geldwäschegesetz (Anti-Money Laundering Act)
H&A	Hauck & Aufhäuser Privatbankiers AG
HNWI	High-Net-Worth-Individual
ICBC	Industrial and Commercial Bank of China
IRA	Individual Retirement Account
IT	Information Technology
KAGB	German Capital Investment Act
KWG	Kreditwesengesetz (Banking Act)
KYC	Know your customer (aka client)
MBTI	Myers-Briggs Type Indicator
MiFID	Markets in Financial Instruments Directive
MiFIR	Markets in Financial Instruments Regulation
MVO	Mean Variance Optimization
MVP	Minimum Viable Product
PNN	Probabilistic Neural Network
PRIIP	Packaged Retail Investment and Insurance-based Products
PSD2	Payment Services Directive 2
REST	Representational State Transfer
S&P	Standard and Poor's
SaaS	Software as a Service
SEC	Securities and Exchange Commission
SOAP	Simple Object Access Protocol
SSL	Secure Sockets Layer
TER	Total Expense Ratio
TLS	Transport Layer Security
UCITS	Undertakings for the Collective Investment in Transferable Securities
UHNWI	Ultra High-Net-Worth Individual
USP	Unique Selling Proposition

VaR	Value at Risk
VAT	Value Added Tax
VermAnlG	German Investment Products Act
WpHG	Wertpapierhandelsgesetz (Securities Trading Act)
WpPG	German Securities Prospectus Act

List of Figures

List of Tables

PART I

Status Quo of Robo-Advisory

CHAPTER 1

Robo-Advisory: The Rise of the Investment Machines

Peter Scholz and Michael Tertilt

Contents

P. Scholz (✉)
Professor for Banking and Financial Markets, Hamburg School of Business Administration, Hamburg, Germany
e-mail: peter.scholz@hsba.de

M. Tertilt
Research Affiliate, Hamburg School of Business Administration, Hamburg, Germany
e-mail: mtertilt@gmx.de

© The Author(s) 2021

P. Scholz (ed.), *Robo-Advisory*, Palgrave Studies in Financial Services Technology, https://doi.org/10.1007/978-3-030-40818-3_1

1.1 Introduction

The Rise of the Machines—these words sound more like the title of a movie or a video game rather than being related to finance. But more than 30 years ago, the story was on the cover of *TIME* magazine:[1] For the first time, the idea that robots and digital solutions may take over the economy was discussed in the public domain. Meanwhile, the digital revolution has disrupted many sectors like the music and publishing industry, imaging technology, retail business, and so on. And since the financial crisis of 2008, the financial sector has got increasingly into the crosshairs of disruptors.

It is noteworthy, however, that the use of technology is nothing unusual for the financial sector. Capital has always been creative through the application of innovations to capture individual advantage. For example, one of the first applications of the telegraph was the long-distance transmission of news which was relevant for trading in distant marketplaces. The invention of the smartphone has been a comparable technical revolution that has had a huge impact on consumer behavior and on the expectations toward banking services. Therefore, it should not come as a surprise that FinTechs came up with the idea to digitalize the advisory process in asset management.

1.2 Robo-Advisory as Part of the Wealth Management Process

The term "robo-advisor" is a blend of the two words *"robot"* and *"advisor"*. It basically describes that the client-interface of the advisory process is not a human anymore but a machine. Based on the wealth management process, in our version adapted from Evensky et al. (2011) (see Fig. 1.1), the question is, which exact process steps are covered by the robo-advisory? In contrast to common perception, robo-advice does not necessarily mean that all these process steps are fully automatized. In fact, the advisory contribution of the robot may be quite diverse. A very common feature of the robo-advisory is the onboarding process of new customers and the determination of the client's risk profile. The management of the sample portfolios, however, is not necessarily a part of robo-advice.

[1] TIME Magazine, May 30, 1983.

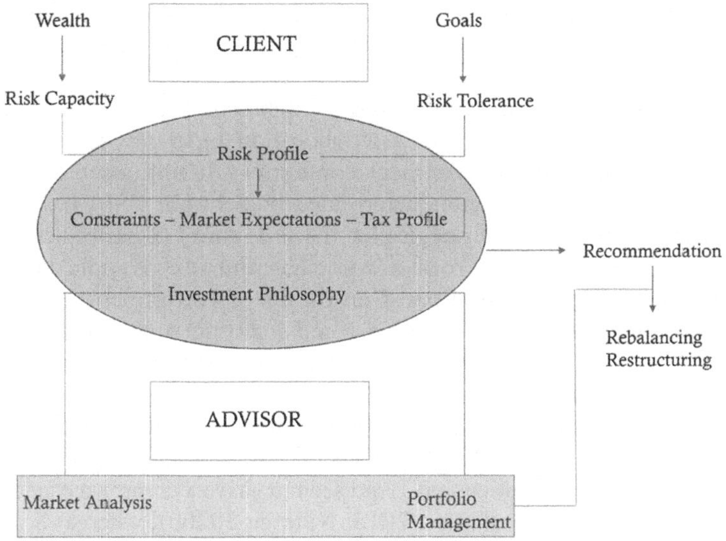

Fig. 1.1 Wealth management process adapted from Evensky et al. (2011)

If we take the largest robo-advisor, Vanguard Personal Advisor Services, as an example, we find a description explaining that the "proprietary algorithm uses [the clients' investment profile] data to recommend a particular investing track" (Vanguard Advisers, Inc. 2019, p.8). So, there seems to be a machine that matches customer data to a specific portfolio. However, there is always a human advisor as well if the client feels the need for any discussion (Vanguard Advisers, Inc. 2019). The dominant robo-advisor, therefore, relies more on a hybrid form of advisory. If we dive deeper into the wealth management process, we find that model portfolios are not generated by a robot but an investment committee: "We may propose the addition, removal, or adjustment of sub-asset class exposures based on continuing portfolio construction research performed by Vanguard Investment Strategy Group" (Vanguard Advisers, Inc. 2019, p. 8). On the other hand, there seems to be an algorithm managing risk: "When recommending, setting, and adjusting your asset allocation, we weigh shortfall risk—the possibility that a financial plan or Portfolio will fail to meet longer-term financial goals—against market risk" (Vanguard Advisers, Inc. 2019, p.8). But there is no active market timing for investments: "The algorithms don't consider prevailing market conditions when making recommendations to you" (Vanguard Advisers, Inc. 2019, p. 10).

If we consider the pursuer, Schwab Intelligent Portfolios, they seem to follow a similar process. The asset allocation apparently is not created by an algorithm but an investment committee: "Using asset allocations and ETF selection parameters determined by Schwab, CSIA has created a number of investment strategies for the Programs" (Charles Schwab & Co., Inc. 2019, p. 2).[2] The robo-advisor algorithm itself is applied to select a portfolio recommendation based on the client's investment preferences for rebalancing, for tax optimization, and for triggering orders. In contrast to the two market leaders, the first robo-advisors, Betterment and Wealthfront, seem to rely more on rule-based allocations. Both explicitly refer to an asset allocation process that is driven inside the robo-advisor and relies on the Modern Portfolio Theory (Wealthfront 2020a; Grealish 2019). In Europe, Nutmeg and Liqid seem to have a committee approach for asset allocation as well (Port 2013; Nutmeg 2020b), whereas Scalable, comdirect, and Quirion operate an explicit asset allocation process as part of their robo-advisory services. A rather lean approach is the service of WeltInvest (Weltsparen 2020), where only sample portfolios with a fixed asset allocation are presented and the investor must choose between the four ETF portfolios. However, there is no advice on which portfolio to select or how to identify risk tolerance, risk budget, and so on. Hence, the investor is basically left on his own. In return, the service is rather competitive with an approximate 0.5% p.a. fee, including product costs.

In summary, it is a bit difficult to describe what really defines a robo-advisor. From our perspective, the very core is the advisory process—the risk profile of the investor is determined, and a portfolio recommendation is given. Hence, we would not denote the offer from WeltInvest as robo-advisor. Within the group of robo-investors, there could be many subgroups. At least, we would form a group based on those robos with a quantitative, rule-based allocation process, and another group that relies on investment committees or external advisors. In conclusion, based on Peter's experience as an investment advisor, the essence of robo-advisory is the replacement of the human advisor, or maybe better, the human salesperson, with an online interface by granting benefits like 24/7

[2] *CSIA* is an affiliate of Schwab, the Charles Schwab Investment Advisory, Inc.

availability, potentially lower costs, scaling, and automatization of at least some aspects of the portfolio management like rebalancing. Consequently, a robo-advisor just substitutes and eases certain steps within the asset management process, but it certainly does not create a whole new process on its own.

1.3 A Brief History of Robo-Advisory

The first robo-advisory services, which were designed for retail investors, came into life in the US after the financial crisis of 2008, probably also as an answer to the increasing distrust in the investment advisory business (Becchi et al. 2018). Betterment, founded by a team including Jon Stein, and Wealthfront, launched by Andy Rachleff and Dan Carroll, are considered as the first robo-advisors in the retail market (Fisch et al. 2018). Only a few years later, the first robo-advisory solutions came up in Europe: The first one founded in the UK was Nutmeg in 2011 (Lielacher 2016), and Quirion was established in Germany in 2013 (Kümpel 2016). Interestingly, the roots of robo-advisory can be traced back either to classic start-ups or to asset managers, whereas the banks by and large did not belong to the early adaptors. Especially for asset managers, robo-advisory services offer a huge leveraging potential (Becchi et al. 2018)—they can easily shift parts of their clients' capital to digital advisory services and lower the entry barrier for new customers. It comes as no surprise that currently the largest robo-advisor, measured in assets-under-management, is owned by Vanguard. The initial idea of robo-advisory has been to disrupt the human advisory services of banks by offering affordable investment advice to widen the customer base by including retail clients and help them make good investment decisions (Nguyen 2018). Today, we can see that robo-advisory is mostly cooperating with financial institutions like banks instead of disrupting them. Based on institutional economics in the FinTech sector (Scholz 2018), it can be shown that mainly due to regulation and trust issues, such as moral hazards, investors are hesitant: What does the robo-advisor do with the money the moment it is invested? Despite the financial crisis, banks still enjoy a kind of bonus trust over start-ups. Moreover, banks are trying to mitigate the distrust issue regarding innovation by applying hybrid models where humans and robo-advisors are combined.

While the early business models have been rather simple—they were concerned with determining the risk preferences of the client and recommending an ETF-based portfolio of stocks and bonds—the advisory tends to become more complex over time, for example, by including more asset classes, active funds, tax optimization, and so on. However, based on our analysis from 2018, we found that the investment process by robo-advisors has been very simple at that time (Tertilt and Scholz 2018). The measurement of risk preferences typically relied only on a few questions, but it does not seem enough to provide proper risk classification and hence to derive adequate investment advice. From that perspective, we assume that the robo-advisory business is still in fledgling stages and further development seems absolutely necessary.

1.4 Clients and Market for Robo-Advice

Since its early beginning, the market volume of robo-advisory has grown significantly. However, it is still small with respect to the global investment industry. The 2018 global assets-under-management (AuM) of robo-advice has been valued at around $ 500 billion (Statista 2020c). To put this number into perspective, the global market for ETFs is ca. nine times larger and accounts for approximately $ 4.7 trillion (Statista 2020a). As one could expect, the US as the country of origin is the dominant market for robo-advice with approximately $ 377 billion AuM in 2018 (Statista 2020b). The biggest player by far is currently *Vanguard Personal Advisor Services*, which belongs to the Vanguard Group, a privately held investment management company. According to Willis Towers Watson (2018), Vanguard is the second largest asset manager worldwide. This seems to confirm the potential leveraging by shifting parts of clients to digital advisory services. Based on their overall AuM of ca. $ 5.2 trillion, Vanguard (2020a) only needs to put a small portion of approximately $ 115 billion (Friedberg 2019) in their robo-advisors to place first. That corresponds to a mere 2.2% of their total AuM. The next rival is *Schwab Intelligent Portfolios*, the robo-advisor of the Charles Schwab Corporation, a large stockbroker company. Compared to Vanguard's robo-advisor, they only manage about a third of the assets with $ 33.4 billion (Friedberg 2019), demonstrating the dominance in market shares by the market leader. This picture becomes even clearer if we consider the two first robo-advisors on the market, *Betterment* and *Wealthfront*, which are comparably dwarfed with ca. $ 16 billion and ca. $ 11 billion, respectively, even though they are ranked third and fourth, respectively, in the list of the US AuM.

The European markets for robo-advice are clearly smaller and just account for around $ 16 billion. Within the European market, the UK (ca. $ 6.1 billion) and Germany (ca. $ 4.5 billion) had the largest market share in 2018 (Statista 2017).[3] Since not all robo-advisors publish reports on a regular basis and most of them have grown at a rapid pace, it is a bit difficult to find current and reliable numbers. Recently, *Nutmeg* is supposed to be the largest robo-advisor in the UK with approximately £ 1.5 billion in AuM (Jones 2019). In Germany, *Scalable*, which is affiliated with Blackrock, is the only robo to hit the € 1 billion barrier (Scalable Capital 2020b) and is now estimated to manage around € 2 billion (Scalable Capital 2020c). These numbers show that the European market still has large potential for development: the estimated growth until 2023 is ca. 41% p.a. This number is in line with the growth of the German market; for the UK market, the estimated growth is ca. 38% p.a. and hence a bit smaller than the European average. On a global level, the growth is estimated to be around 31% p.a., which is also the result of the comparatively slow expansion in the largest market. The US is only estimated to increase the AuM by ca. 25% p.a. The largest driver of global growth is supposed to be China. With an approximately AuM of $90 billion in 2018, the Chinese market is estimated to increase to $ 880 billion in 2023, which corresponds to an annual growth rate of ca. 46% p.a. (all growth estimates calculated with Statista (2017) data). However, at least for 2023, the US is still seen as being the largest market for robo-advice based on Statista (2017) forecasts.

Breaking down from markets to customers, around 6.6 million clients currently use robo-advisory services in the US. This number is supposed to grow by ca. 15% p.a. until 2023. Looking at the per capita investment, an average US robo-advice investor in 2018 put approximately $ 65,000 on his or her account. This number is assumed to grow quickly until 2020 to a bit more than $ 100,000 and then levels there. In China, the second biggest market in terms of AuM, around 17.5 million people used robo-advice in 2018. And this number is estimated to grow by ca. 39% p.a. to ca. 124.5 million people in 2023. The per capita investment, however, is significantly smaller than in the US: the average Chinese investor in 2018 placed only around $ 5100 in robo-advice. And this number probably increases only

[3] The numbers in the Statista dossier include data until end of 2017 and forecasts for 2018 and beyond. To have a fair standard of comparison, the 2018 numbers are used which may include forecasts sometimes.

by ca. 7% p.a. over time. The clear driver in China seems to be the number of clients and not the average investment amount. In Europe, there are currently only around 900,000 robo investors, but the forecast sees a ca. 28% p.a. expansion. Regarding the per capita investment, the Europeans are in-between the US and China: currently, on average, ca. € 15,500 are placed on an account with a growth perspective of about 13% p.a., which gives an average investment of ca. €29,000 in 2023. Since we do not have exact information about how the investments are distributed among investors, the interpretation of these numbers gives some leeway. One potential explanation could be that especially in Europe and China, robo-advice is indeed used for retail investors that use the digital service to benefit from professional investment advice that they could not afford otherwise. In the US, by contrast, middle-class investors seem to use robo services. Another explanation attempt points to a trust issue. Maybe investors in Europe and China do not fully trust the robo-advice and therefore use only a smaller portion of their wealth; they also distribute their investments among different suppliers and use human advisors as well. This could be especially true for Europe, which is often seen as more skeptical about digitalization. At least for Germany and Italy, this assumption is backed by two studies of Kaya (2017) and Kaya (2019), which find that the average investor possesses medium or high wealth (result for Italy), has around €4000 as monthly net income, robo-invests between € 1000 and € 1500 annually (Germany), and typically has a university degree (both countries). At least, this points more toward private banking than a retail banking clientele. In China, by contrast, people tend to be very technology-friendly, but maybe the trust issue here points more at the direction of institutions—this may explain the lower investment amounts. In the US, people tend to be more open-minded regarding technology and hence may rely more easily on the robo-advisor than on the bank (all forecasts in this paragraph are based on Statista (2017) data).

It seems to be a natural assumption that the typical investors of robo-advice are the millennials who are aged 24 to 35 years (Zeldis Research Associates 2018). However, according to Zeldis Research Associates (2018), millennials do not share exuberance for robo-advisory, but they have "limited awareness" of these services. Kaya (2017) states that the millennials widely belonged to the early adopters, but this has shifted significantly. Based on Kaya (2017) and Kaya (2019) statistics, the age of the average investor is rather between 45 and 48 years—it holds for both US and German investors. Moreover, based on estimates from Germany, the overwhelming number of investors is male (Kaya 2019). In Italy, the distribution is a bit more balanced, but there are still significantly more male than female investors applying digital investment advice (Kaya 2017). The dominance of male investors is probably also a fair assumption for most other robo-advisory markets.

Despite the relatively optimistic development predictions, trees do not grow in the sky in the robo-advisory business as well. To be profitable, the provider of robo-services needs to raise their AuM above a certain threshold (Jones 2019). Currently, there are more than 200 robo-advisors in the US, 30 in Germany, and 20 in the UK. More and more financial institutions try to establish a robo-advisor, maybe as an open platform or solely as a distribution channel for their own products—for example, ICBC, one of the largest Chinese banks, has recently tried to launch a robo-advisor for their clients. In a report from March, HSBC estimates that a robo service in North America, which charges 0.25% as fee, needs approximately between $ 11.3 billion and $ 21.5 billion AuM to break even (Jones 2019).[4] At present, only the top four providers in the US reach these numbers. In Europe, where HSBC observed higher fees at around 0.45%, at least more than $ 3.5 billion AuM seems to be necessary. Given that the scope for increasing fees seems to be limited, most robo-advisors need to collect more money and acquire more clients to become profitable. Hence, a consolidation of the sector seems to be inevitable. At least in Germany, some well-known financial institutions have abandoned their plans with robo-advisory—for example, Commerzbank, the second largest bank in Germany, stopped their plans launching its own robo-advisor and now relies on the service of their affiliate comdirect. A similar case is the Spanish bank Santander, which recently stopped their robo-advisor. While banks typically have a customer base, which they could convince of the robo-advice, for start-ups it is harder to gain trust and attract more clients. Therefore, a cooperation between FinTechs and banks seems to be a reasonable way to address consolidation. While there seems still to be plenty of room for innovative providers with reasonable fees, it will play a key role in matching customer needs and expectations with respect to performance and quality of advice. However, for those asset managers who do not have a clear and precise philosophy, or those who charge rather high fees, it will probably become a tough market.

[4] Original study of HSBC could not be obtained.

1.5 Performance of Robo-Advisors

To create a competitive service, it comes back to the performance at the end. Irrespective of the elaboration of the service, if the (risk-adjusted) returns do not convince, the robo-advisor will not be successful in the long run. To measure the historical performance of robo-advisors can be a little tricky, since, to the best of our knowledge, there is no need to publish data nor to create a database to collect the performance of different robo-advisors internationally. Hence, we can only provide some statistics with varying granularity and quality of data. A rather current development is AISP platforms, which allow for individual comparisons between different robo-advisors.

The largest robo-advisor from Vanguard does not seem to publish detailed historical performance data. It states, however, that approximately "85% of our funds have performed better than their peer-group averages over the last 10 years" (Vanguard 2020b). Moreover, on a website that compares different robo-advisors, we found information that "Vanguard' claims its US stock funds offer a 7.51% rate of return in line with S&P 500 avg returns" (Moore 2019). So, generously spoken, an investor seems to earn approximately market returns while investing in Vanguard's robo-advisor. It remains unclear if this holds before or after fees. For Schwab's robo-advisor, the situation seems to be pretty similar: Schwab does not publish charts or return data but a quarterly written report (Charles Schwab & Co., Inc. 2019). In the section "How did Schwab Intelligent Portfolios do?", the management gives brief information about the development of the different strategies in a rather qualitative way—there are hardly any numbers presented. Wealthfront at least provides some return numbers for their different sample portfolios (Wealthfront 2020b). Even if the information is not very granular, Wealthfront states returns based on different frequencies: annual, three years, five years, and since inception. This allows the investor to get at least a certain feeling about the past performance. However, the risk figures are not displayed. Looking at the tax-optimized portfolio with the highest risk score, Wealthfront has yielded 8.05% p.a. since inception (December 15, 2011, until July 31, 2019). This is significantly lower than an investment in the S&P 500 index, which would have earned an annual 14.4% if dividends had been reinvested (DQYDJ 2020). Even if we consider that this number does not include fees for an ETF or tax abbreviations, it seems to be significantly higher than any robo-advice-based investment with Wealthfront at first glance.

Still, we should keep in mind that the Wealthfront portfolio is supposedly better diversified and may therefore carry less risk. This comparison can only deliver a first indication about the "real performance" of the robo-advisor. The most detailed performance information from the big four US robo-advisors is provided by Betterment: they display a chart with monthly returns from their different portfolio strategies which dates back to January 2004, well before their official launch (Egan 2014). If we take a similar investment period (here: December 2011 until February 2019), a 100% stock portfolio yielded an annual return of ca. 9.64%. During the same period, the S&P 500 increased by ca. 14% p.a., if one includes dividends, but before fees and taxation. The gap seems to be smaller for Betterment than for Wealthfront. But again, since we have only little information other than returns, the comparison is limited in its value. Another source of performance information is Statista (2019), which claims to provide a global robo-advisory comparison for 2018 based on a report by BackendBenchmarking (2018). In their analysis, however, all the robo-advisors are based in the US. In this ranking, the largest robo-advisor from Vanguard is placed first with a 2018 return of 6.56%, followed by WiseBanyan (6.25%). Other robos with large AuM are ranked #7th (Schwab with 4.77%) and #8th (Betterment with 4.17%). In the Statista analysis, it is not clearly stated which portfolios they exactly use or how they exactly perform the analysis. Hence, this should also be taken with a grain of salt.

In the UK, Nutmeg provides charts for their different portfolios: monthly returns, which can be seen in the chart, daily returns, which solely are given by the chart, and an average competitor comparison based on Asset Risk Consultants (ARC) data (Nutmeg 2020a). They declare their returns after fees and with reinvested dividend payments. Their "all-time" data contains the period from September 30, 2012, until July 31, 2019, and reports a 9.2% annual return for the portfolio with the highest risk compared to the ARC average competitor return of 8.4% p.a. Looking at the calendar years, Nutmeg outperformed the average in each year with positive returns. In 2018, however, the first year with negative returns of 9.9%, they failed to beat their benchmark (−6.5%). Also, in the current year, they are behind their benchmark (2.4% compared to 5.4%). So, it seems that they have fallen behind the average during the past one and a half years. Trying to analyze the chart, the maximum drawdown happened in mid-2018 and recovered until April 2019 with an approximate 25% downturn, which seems a bit high for the period of evaluation. Since the equity portfolio of Nutmeg has rather high stakes in both US (ca. 43%) and UK (25%) stocks, we look at the FTSE and S&P500 performance as a standard of comparison. For the S&P500, the performance shows an annual return of 13.38%; for the FTSE, it is 7.58%. If we weight the returns of both indices accordingly, we receive an annual return of around 11.25%, which is comparable to the 9.2% return of Nutmeg, keeping in mind that index data does not contain fees and is not as diversified as the Nutmeg portfolio. Comparing Nutmeg to its peers, there is a UK website providing performance data for different suppliers from July 1, 2017, to June 30, 2018 (Boring Money Business 2020). For the high-risk portfolios, the Nutmeg 10 portfolio showed the highest return (7.92%), although it falls behind the FTSE benchmark (8.4%). The other robs-advisors followed within a range of 7.46% (IG 5 Aggressive) to 4.15% (Scalable 25% VaR). The question remains: how they would compare in risk-adjusted returns as the level of diversification is probably not the same. Still it is an interesting result to compare the highest risk portfolios of different robotic managers, assuming that the suppliers offering the highest returns probably also follow the most aggressive investment styles.

In Germany, robo-advisors tend to publish only little performance information. The market leader in AuM, Scalable, displays a graph of different value-at-risk strategies that they offer (Scalable Capital 2020a). It is important to note that even the most aggressive strategies are not 100% invested in equities but in other asset classes like bonds, real estate, and so on. Hence, they compare to a mixed-funds benchmark from Morgan Stanley. Since their inception in January 2016, they show similar returns. However, the deflections and hence the risk seem to be somewhat lower for the robo-advisor than for the mixed-fund. Liqid also provides a chart for their performance, which is not very granular (Liqid 2020). The riskiest portfolio has currently an equity quota of around 95%. Liqid uses the average of German wealth managers as the benchmark. Since April 2016, they performed significantly higher than their benchmark (ca. 33% vs. 15.5% in total). Quirion delivers a table with annual performance information (from June to June of the corresponding years), but they do not compare to a benchmark (Quirion 2020). In their 100% equity strategy, Quirion achieved rather high returns in four years (19.9%, 16.4%, 15.8%, and 6.6%) and a negative return in one year (−5.6%). Owing to the granularity of data, no risk information is available. For the robo-advisor of comdirect, no public available performance information was found. For the German market, we found a website that tests the robo-advisors with real money and publishes the performance data (Broker Vergleich 2020).[5] In this test, comdirect yields the highest returns in 2019, followed by smaller robo-advisors. The next big player is Scalable, which is ranked seventh.

We would not like to interpret too much into these results. However, it becomes clear that portfolios that are made for a certain type of investor can show a wide range of different return and probably risk characteristics. Robo-services as rational agents do not seem as homogenous as one might expect. Hence, it seems to be a wise recommendation to check carefully the information provided by suppliers while exploring the offers of different vendors. Another key takeaway is our impression that by and large the performance of robo-advisors seems to be reasonable and that most of them do a decent job of managing the money of their clients. The ongoing Corona crash is most certainly a true stress test for robo-advisors. It will be interesting to learn which robo-advisor philosophy—algorithms or committees—will be more successful in managing the extremely high volatilities in the market.

1.6 Robo-Advice in a Nutshell

As of today, banks and asset managers widely see robo-advisory as a promising new sales channel for investment management. Depending on the specific robo-solution, the capital requirements to access these services can be rather low, at least lower than in the traditional wealth management industry. Hence, it is tempting for many established asset managers to introduce robo-advisory to select clients between different segments. However, it would be a pity if robo-advisory would be interpreted only as a low-budget solution for the "not so rich" investors in the future. Maybe, the true potential lies in the integration of both human and robo-advice, which is also known as *hybrid model*. Some tasks like data collection most certainly can be done more efficiently by a robo-advisor. In other aspects, such as displaying empathy in response to the individual client's situation, the human is superior. At least, this is currently especially the case for the matter of trust. By and large, robo-advisory emerged after 2008 and hence did not have to face a real bear market so far. It will be interesting to see how robo-advisors would perform in a plummeting market. It is an essential question if purely digital advice and online interfaces would

[5] Another website that compares information from robo-advisors is biallo.de (2020).

be enough in such a scenario to cover the larger information need of clients and maybe even to provide emotional support them. Undoubtedly, these requirements of human investors will have their impact on the further development of robo-advisory services and may accelerate the tendency to hybrid approaches. Another important factor will be the issue of automatization within the wealth management process itself. Robo-advice is predestined for rule-based investments and algorithms watching over portfolios for rebalancing and risk management purposes. And finally, technologies like predictive banking and artificial intelligence may find their way into robo-advisory. So, there is plenty of room for future development. In fact, it will be interesting to see which innovations and interpretations of the automated service will thrive the advisory business in the near future.

References

BackendBenchmarking. 2018. *The Robo Report. Third Quarter 2018.* Last Accessed: March 20, 2020. https://storage.googleapis.com/gcs.backendb.com/1/2019/04/2018_3Q.pdf.

Scott M. Becchi, Ugur Hamaloglu, Taroon Aggarwal, and Samit Panchal. 2018. *The Evolution of Robo-advisors and Advisor 2.0 Model: The Future of Investment Management and Financial Advisory.*

biallo.de. 2020. *Die Besten Robo-Advisor im Performance-Vergleich.* Fonds. Last Accessed: March 20, 2020. https://new.biallo.de/vergleiche/etf-fonds-aktien/robo-advisor/.

Boring Money Business. 2020. *Robo Adviser Performance—12 Months to 30th June 2018.* Last Accessed: March 20, 2020. https://www.boringmoneybusiness.co.uk/research/robo-report-q2-2018/.

Broker Vergleich. 2020. *Daten zur Performance aller Robo Advisors im Test.* Online-Broker. Last Updated: March 31, 2020. https://www.brokervergleich.de/robo-advisor/echtgeld-test/.

Charles Schwab & Co. Inc. 2019. *Schwab Intelligent Portfolios and Q2 2019 Market Performance.* Last Accessed: March 20, 2020. https://intelligent.schwab.com/public/intelligent/insights/blog/q22019-performance-review.html.

Charles Schwab & Co., Inc. 2019. *Schwab Intelligent Portfolios Solutions, Disclosure Brochure.* Last Accessed: September 29, 2019. https://www.schwab.com/public/file/CMS-BDL100049.

DQYDJ. 2020. *S&P 500 Return Calculator with Dividend Reinvestment.* Last Accessed: March 20, 2020. https://dqydj.com/sp-500-return-calculator/.

Dan Egan. 2014. *Explore Betterment's Historical Performance.* Last Accessed: March 20, 2020. https://www.betterment.com/resources/betterment-historical-performance/.

Harold Evensky Stephen M. Horan, and Thomas R. Robinson. 2011. *The New Wealth Management: The Financial Advisor's Guide to Managing and Investing Client Assets*. CFA Institute Investment Series. Wiley.

Jill E. Fisch, Marion Laboure, and John A. Turner. 2018. "The Emergence of the Robo-Advisor". *Wharton University of Pennsylvania Pension Research Council*. Working Paper WP2018-12. https://repository.upenn.edu/prc_papers/10/.

Barbara A. Friedberg. 2019. *Robo-Advisors With the Most Assets Under Management—2019*. Last Accessed: March 20, 2020. https://www.roboadvisorpros.com/robo-advisors-with-most-aum-assets-under-management/.

Adam Grealish. 2019. *The Betterment Portfolio Strategy*. Last Accessed: March 20, 2020. https://www.betterment.com/resources/betterment-portfolio-strategy/.

Alexander Jones. 2019. *Why Robo-Advisors are Struggling to Break Even*. Last Accessed: March 20, 2020. https://internationalbanker.com/brokerage/why-robo-advisors-are-struggling-to-break-even/.

Orçun Kaya. 2019. *German Robo-Advisors: Rapid Growth, Robust Performance, High Cost*. Deutsche Bank Research. https://www.dbresearch.com/PROD/RPS_EN-PROD/PROD0000000000487351/German_robo-advisors%3A_Rapid_growth%2C_robust_perform.pdf.

Orçun Kaya. 2017. *Robo-Advice—A True Innovation in Asset Management*. Deutsche Bank Research. https://wwwdbresearchcom/PROD/RPS_ENPROD/PROD0000000000449125/Roboadvice_%E2%8%93_a_true_innovation_in_asset_managemen.pdf.

Anna-Lena Kümpel. 2016. *Fintech: Bei der Geldanlage geht der Trend zum Robo Advisor*. Last Accessed: March 20, 2020. https://berlinvalley.com/fintech-geldanlage-trend-robo-advisor/.

Alex Lielacher. 2016. *The Complete Guide To UK Robo Advisors*. Last Accessed: March 20, 2020. https://smartmoneysmartliving.com/2016/10/12/complete-guide-uk-robo-advisors/.

Liqid. 2020. *LIQID Wealth*. Last Accessed: March 20, 2020. https://www.liqid.de/liqid-wealth.

Barry D. Moore. 2019. *What Is A Robo Advisor? Hybrid vs AI vs Robo Advisors*. Last Accessed: March 20, 2020. https://www.liberatedstocktrader.com/what-is-a-robo-advisor/.

Phuong Nguyen. 2018. *Betterment: Jon Stein*. Last Accessed: March 20, 2020. https://www.npr.org/2018/1/26/661049018/betterment-jon-stein?t=1564478275377&t=1584701835665.

Nutmeg. 2020. *Fully Managed Portfolios*. Last Accessed: March 20, 2020. https://www.nutmeg.com/fully-managed-portfolios.

Nutmeg. 2020. *How We Invest*. Last Accessed: March 20, 2020. https://www.nutmeg.com/how-we-invest#portfolio-performance.

Shaun Port. 2013. *An Inside View of the Nutmeg Investment Process: Asset Allocation*. Last Accessed: March 20, 2020. https://www.nutmeg.com/nutmegonomics/an-inside-view-of-the-nutmeg-investment-process/.
Quirion. 2020. *Über Uns*. Last Accessed: March 20, 2020. https://www.quirion.de/geldanlage/.
Scalable Capital. 2020. *Erzielte Rendite*. Last Accessed: March 20, 2020. https://de.scalable.capital/erzielte-rendite.
Scalable Capital. 2020. *Robo-Advisor*. Last Accessed: March 20, 2020. https://de.scalable.capital.
Scalable Capital. 2020. *Über Uns*. Last Accessed: March 20, 2020. https://de.scalable.capital/ueber-uns.
Peter Scholz. 2018. "Hostile Takeovers: Are Banks Threatened by FinTechs?" *Corporate Finance* 01-02: 3–7.
Statista. 2020. *Entwicklung des weltweit in ETFs verwalteten Vermögens von 2005 bis 2019*. Last Accessed: March 20, 2020. https://de.statista.com/statistik/daten/studie/219372/umfrage/weltweit-in-etfs-verwaltetes-vermoegen-seit-1997/.
Statista. 2019. *Performance of Selected Robo-Advisors 2018*. Last Accessed: September 29, 2019. https://www.statista.com/statistics/755599/performance-of-selected-robo-advisors/.
Statista. 2020. *Prognose zur Entwicklung des verwalteten Vermögens der Robo-Advisors in den USA von 2017 bis 2023*. Last Accessed: March 20, 2020. https://destatista.com/statistik/daten/studie/740573/umfrage/entwicklung-des-verwalteten-vermoegens-der-robo-advisors-in-den-usa/.
Statista. 2017. *Robo-Advice: Dossier*. https://de.statista.com/statistik/studie/id/48179/dokument/robo-advice/.
Statista. 2020. *Robo-Advisors. Worldwide*. Last Accessed: March 20, 2020. https://www.statista.com/outlook/337/100/robo-advisors/worldwide.
Michael Tertilt and Peter Scholz. 2018. "To Advise, or Not to Advise—How Robo-Advisors Evaluate the Risk Preferences of Private Investors". *The Journal of Wealth Management* 21(2): 70–84.
Vanguard. 2020. *Fast Facts About Vanguard*. Last Accessed: March 20, 2020. https://about.vanguard.com/who-we-are/fast-facts/.
Vanguard. 2020. *Why Vanguard*. Last Accessed: March 20, 2020. https://investor.vanguard.com/why-vanguard.
Vanguard Advisers, Inc. 2019. *Vanguard Personal Advisor Services Brochure*. Last Accessed: September 29, 2019. https://personal.vanguard.com/pdf/vpabroc.pdf.
Wealthfront. 2020. *Wealthfront Investment Methodology White Paper*. Last Accessed: March 20, 2020. https://research.wealthfront.com/whitepapers/investment-methodology/.

Wealthfront. 2020. *Wealthfront's Average Annual Net-of-Fees, Pre-tax Returns.* Our Historical Returns. Last Updated: March 31, 2020. https://www.wealthfront.com/historical-performance.

Weltsparen. 2020. *WeltInvest. Das ETFPortfolio für clevere Anleger.* Last Accessed: March 20, 2020. https://www.weltsparen.de/etf-portfolio/.

Willis Towers Watson. 2018. *The World's Largest 500 Asset Managers. Thinking Ahead Institute and Pensions & Investments Joint Research.* Last Accessed: March 20, 2020. https://willistowerswatson.pl/aktualnosci/pdf/2018.10.29-TAI-PI-500_2018.pdf.

Zeldis Research Associates. 2018. *Uncertain Futures: 7 Myths about Millennials and Investing.* Tech. rep. Last Accessed: September 29, 2019. FINRA Investor Education Foundation and CFA Institute. https://www.cfainstitute.org/en/research/survey-reports/millennials-and-markets-2018.

CHAPTER 2

Situating Robo-Advisory

Sinan Krueckeberg

Contents

2.1 Introduction

Academic research into robo-advisory is still nascent but developing dynamically, especially in light of an increasing amount of attention directed toward what is doubtlessly one of the 'hot-topics' in the field of finance today. Inquiry can to date be clustered into four distinct streams that are sequentially linked to each other: (1) A definitorial cluster that aims to situate robo-advisory in the broader context of the financial industry in general and financial technology in particular. (2) Analyses of strengths and weaknesses of robo-advisory, especially in terms of differentiation

S. Krueckeberg (✉)
Krueckeberg Family Office, Hamburg, Germany
e-mail: sinan@krueckeberg.us

© The Author(s) 2021

P. Scholz (ed.), *Robo-Advisory*, Palgrave Studies in Financial Services Technology, https://doi.org/10.1007/978-3-030-40818-3_2

from each other and from legacy products and services. (3) Building on analyses regarding strengths and weaknesses, a third strand of research asks the question of what impact robo-advisory will thus have on incumbents. (4) The fourth strand of inquiry, finally, aims to shed light onto the path ahead for this emerging field. This chapter will synthesize and discuss extant literature emanating from academic research into robo-advisory and cluster key insights, according to the four streams of inquiry outlined above, within distinct sections.

2.2 What Are Fintech and Robo-Advisory, Really?

Puschmann (2017) describes financial technologies as offshoots of an ongoing macro process of economy-wide digitalization. The reasons why the process of digitalization through the implementation of information technology is developing a pronounced force within the realm of financial services is seen in two characteristics. Firstly, in that financial products are intrinsically based on information which, through the use of information technology, can be channeled and used in countless new fashions. Secondly, in that most processes within the financial industry can be implemented almost entirely without physical contact. This latter argument regarding physical interaction is particularly interesting within the scope of this book, as the field of advisory presents a unique exception—a phenomenon that will be discussed in many of the following chapters. The spread of information technology within the financial industry is expected to lead not only to the automation of certain elements of the value chain, but to a fundamental reorganization of entire chains altogether. This reorganization is observed to be driven by two dynamics: on the one hand, by the market entry of new actors into financial services that are digital-native at their core, such as Amazon, Apple and the like; on the other hand, through the introduction of innovative new business models driven by emergent technological solutions, such as robo-advisory. Ferrari (2016) and Puschmann (2017) seem to agree that two of the key drivers behind this trend are (1) changing consumer behavior, showing an increased quantity and quality of interaction of consumers via digital channels, and (2) changing regulation following the subprime crisis that imposed significant financial and compliance burdens onto legacy players, putting them at a disadvantage versus flexible digital challengers. Connecting to

Puschmann, Ferrari (2016) expects the financial industry to transition from a universal banking model to verticalization that allows focused challengers to develop highly specialized niche capabilities in verticals such as credit products, transactions, or robo-advisory. Evidence provided by Dorfleitner et al. (2017) supports this proposition, showing that verticals driven by financial technologies have shown average annual growth rates of 150% over the years 2010 to 2016 in Germany alone. How, then, is the scope of robo-advisory delineated within the academic discourse? Sironi (2016) frames robo-advisory as an information technology–based, automated investment advisory service. Following Maedche et al. (2016), robo-advisors can be classified as based on interactive user-assisted systems that possess a certain degree of context awareness. Taking these elements together, robo-advisors help to create and manage investment accounts on a fully automated basis (Jung et al. 2018). Reflecting on these foundational concepts leads Jung et al. (2018) to conclude that robo-advisory embodies the capability to transform the advisory process from a traditional human-to-human setting into one that can potentially be entirely human-to-machine-based. Key metrics that traditionally were compiled by a human advisor are translated into digital routines for data gathering and translated into investment advice along the typical investment dimensions of risk, return, liquidity, and so on. This is seen as a catalyst for significant cost reductions in advisory services, in effect lowering the barriers to participation for individual investors—robo-advisory as a force toward the democratization of asset management. While certainly an interesting thought, challenging the structure of the established financial order will not be an easy feat. Above all else, clear differentiation of the offering based on leveraging strengths and compensating weaknesses will be a decisive factor for the success or failure of challengers.

2.3 Differentiation: Strengths and Weaknesses

Robo-advisory as a new fintech enabled banking vertical is without doubt an enticing proposition. However, new vertically specialized contenders will need to differentiate themselves both internally from each other as well as externally from traditional advisory to develop a unique value proposition based on their strengths that can provide increased value to customers. Turning first to internal differentiation among robo-advisors,

Sironi (2016) provides a concise overview that arrives at four key factors along which propositions align on the competitive landscape. The first factor is the degree of passive management. Some robo-advisors might choose to manage portfolios entirely passively with a limited preset of broad index products and very infrequent rebalancing by which portfolios are recalibrated only rarely in reaction to movements in the market. Others might take a more active approach that involves both more granular picking of securities and more frequent rebalancing. This connects further to the second element of differentiation, namely that of the depth of investment automation. To what degree are investors involved in the ongoing investment process, if at all? Some robo-advisors might involve customers more actively in the investment process, while others might advertise a hands-off solution. Third, such an investment process itself will be guided and differentiated by some form of assessment mechanism that aims to translate preferences and circumstances of the individual investor into a tailor-made asset allocation. Thereby, precision and fidelity of investor classification could provide a source for product differentiation. This in turn connects to the fourth element, the target clientele. Some robo-advisors might choose to focus on the tech-savvy Millenials that value convenience above all else, others on the affluent baby boomers which might have a different set of preferences and needs altogether. Some might value virtual-only experiences, others require a hybrid approach—a question that will be discussed in detail in Sect. II of this book. Wherever robo-advisors choose to align along these four factors regarding internal differentiation, common characteristics arise that can be isolated as factors for external competition versus traditional advisory.

Analyzing competitive strengths, Jung et al. (2018) point out the cost-competitiveness that digital services in general and robo-advisors in particular can use to their advantage, a point that is driven particularly by strong scalability of the service offering, which can be available 24/7. Sironi (2016) concurs by emphasizing cost-efficiencies on the part of such challengers as a key strength. Without having to finance costly real-estate and an expansive network of advisory staff that is limited in both accessibility and total hours of availability, robo-advisors are found to be able to offer their services both at lower fees and with a lower minimum threshold of invested capital (Jung et al. 2018). Meola (2017) shows that minimum invested capital can be as low as US$ 0. In times of increasing regulation and compliance requirements, customer on-boarding has become a focal point for financial services companies. Sironi (2016) finds advantages on the side of robo-advisors versus incumbents, not least carried by technological innovations such as customer-not-present video identification and intuitive self-profiling. This ties into the point of overall simplicity, which is argued to be supportive for both the customer journey and the overall innovation process in robo-advisory (Sironi 2016). Jung et al. (2018) point out two further strengths that are somewhat linked: automated rebalancing and tax-loss harvesting. Where customers of traditional advisory might either engage themselves in active rebalancing with the assistance of various digital and analogue tools or pay for human advisors to coordinate rebalancing, robo-advisors are seen as reliable executors of predefined granular rebalancing strategies. This ties into the automation of tax-loss harvesting, through which automated algorithms optimize rebalancing with the aim of harvesting advantages based on tax-rate arbitrage.

Turning to current weaknesses, Fein (2015) studies the user agreements and/or disclosure brochures of three leading US robo-advisors in detail and paints a more critical picture of the robo-advisory landscape. A counterpoint to the cost-efficiency argument is supplied in that while customers might not be required to explicitly pay for advisory services consumed as such, costs associated with these services are later recovered and thereby paid for by the consumer through third-party fees such as brokerage costs, transaction fees, and other expenses. This might occur through outright fees, all-in fees, or revenue-sharing arrangements with third parties. This links to a related point of criticism regarding the independence from conflicts of interest on the part of robo-advisors. Fein lays out that within the scope of her study, explicit mention was made on the parts of robo-advisory firms about being tied to affiliated brokers that might not provide best execution of clients' trading orders. The robo-advisor's own profit motive might stand against best execution of its customers. Furthermore, some robo-advisors disclose and request consent for proprietary trading and reserve the right to recommend investments to customers that are being held in proprietary portfolios. This can lead to biased investment recommendations especially paired with a further point of criticism, namely some robo-advisors reserving the right to aggregate and pool client orders as well as to execute in the form of cross transactions. All these points in conjunction do paint a picture of potential for conflicts of interest. Two final points of criticism are that robo-advisors do not truly provide tailor-made investment advice, which is tied to further criticism regarding the assessment of individual risk tolerance and the level of overall personalization of the user experience (Fein 2015). Tertilt and Scholz (2018) connect to Fein's research and supply supporting evidence with an analysis of the robo-advisory market, specifically, how robo-advisors translate individual customers' risk tolerance into equity exposures. Generating a proprietary data sample comprising robo-advisors from Germany, the UK, and the US, Tertilt and Scholz (2018) show that robo-advisors ask only relatively few questions, and, of these, only about 60% have any impact on the actual risk categorization. Customers seem to receive only little guidance in the process of assessing their idiosyncratic risk budget and risk appetite (Tertilt and Scholz 2018). Faloon and Scherer (2017) provide further support for this thesis by summarizing questions typically used by most robo-advisors and thereby concluding that most of the generated advice is generic and does not qualify as personalized advice. In light of current strengths and weaknesses, what can the potential impact of robo-advisory be on legacy players and how can established advisory providers react to challengers?

2.4 Impact on Incumbents

Who are the early adopters making use of these new digital service offerings? Woodyard and Grable (2018) study the characteristics of customer segments that exclusively use either robo-advisory or traditional advisory. As could be expected, they find that individuals who exclusively use robo-advisors as their service provider are commonly young, confident in their own abilities, and distrustful of traditional channels of financial advice. Epperson et al. (2015), based on their study among 4000 US consumers, support these results by reaching a similar conclusion. Whether this constitutes the beginning of a broader trend for future generations to increasingly transition toward purely digital service offerings or whether preferences will shift when today's youth become tomorrow's seniors remains to be seen. Whichever the trajectory will be, incumbents are today confronted with the emergence of competing value propositions. How should incumbents react? Cocca (2016b) studies the impact of virtual advisory models onto traditional advisory, based on a survey among wealth management clients in Austria, Germany, and Switzerland. Considering what he calls the potential virtualization of advisory, Cocca arrives at the conclusion that the competition between traditional advisory incumbents and digital challengers will play out along a complexity versus standardization spectrum. According to this approach, challengers will migrate some service offerings into digital processes that are highly standardized and automated, such as risk-profiling and/or portfolio management. Traditional advisors in turn are expected to be forced into more complex advisory services such as international asset structuring and/or tax advisory (Cocca 2016a). Jung et al. (2018) provide findings that can be connected to Cocca, in that robo-advisors are expected to focus on individuals with below-average to average income as traditional advisory is expected to focus on the upper end of the spectrum. Summarizing both sources, this would mean that robo-advisory might find success in focusing on standardized and automated low-cost services, while traditional advisory might focus on higher-cost/higher-margin complex advisory services. Cocca (2016a) adds to this, concluding that wealth management specifically might find a winning proposition in a hybrid approach. To be able to engage customers with a hybrid strategy, Cocca recommends incumbents not to understand fintechs as competitors but as potential integration partners (Cocca 2016b). Gold and Kursh (2017) concur with the idea of integrating fintechs into incumbent organizations and add three further strategic choices, using D'Aveni's framework of strategic options for incumbents as their reference (D'Aveni 2002). Besides acquiring challengers, incumbents might launch their own digital offerings, leveraging internal resources and thereby building on established customer relationships. Or they might partner with challengers using their value proposition as a complement to traditional services. Finally, incumbents can also choose to match parts of the offering of challengers, that is, by lowering fees, thereby eliminating an important incentive for customers to switch service providers. Overall, Gold and Kursh (2017) conclude that taking a wait-and-see approach might be the most favorable course of action in light of regulatory circumstances and lock-in effects for incumbents due to proprietary customer data. Whether proprietary data in and of itself will remain a competitive factor in the emerging Application Programming Interface (API) economy, which is increasingly based on the free exchange of information, remains to be seen. Gold and Kursh recognize this, especially in light of recent European legislation, not least the Payment Services Directive II (European Parliament and the Council of the European Union 2015), mandating banks to grant fintechs access to their customer data via API.

2.5 The Path Ahead

Which developments and success factors does the academic literature anticipate for the future? Arwas and Soleil (2016) anticipate the next-generation Robo-Advice 2.0 and expect success factors not to lie in technology itself but rather in the creation of personalized and differentiated offerings that are geared toward serving the entirety of the customer life cycle. Among the factors that will play a crucial role in robo-advisory being able to transition into the mainstream over the coming years, Arwas and Soleil (2016) highlight simplicity in both language and process. Moreover, emphasis is put particularly on transparency regarding costs—a thought that can be seen as linking back to Fein's 2015 critique of robo-advisory, as discussed above. Key emphasis on transparency is a thought that Jung et al. (2018) echo with the results of their study. Designing a robo-advisor within a laboratory environment, Jung et al. analyze the requirements of both experts and potential customers for robo-advisory solutions. To do this, first, design principles are extracted from existing literature, which are then, second, validated via laboratory experiments. Four design principles are summarized that have in the past been suggested as fundamental to digital financial advisory success. These are ease of interaction, work efficiency, information processing, and transparency. Sources for these principles are studies conducted by Koerner and Zimmermann (2000), Moewes et al. (2011), Nussbaumer (2012), Nueesch et al. (2014), Kilic et al. (2015), and Ruf et al. (2015). Jung et al. (2018) break down the four design principles into design requirements and thereafter analyze the appropriateness of each factor in the so-called 'design cycles'. The first cycle includes expert interviews, the second and third cycles comprise a mixed-method experimental usability study in the laboratory, where prospective customers are analyzed for their likelihood to adopt the prototypical solution. By doing this, Jung et al. (2018) find evidence that all four design principles indeed are appropriate guiding elements to build future digital advisory services around and that transparency can be identified as the single most important aspect. Regarding the specific variables to structure the advisory process itself around, Scherer (2017) conducts a panel survey with German households and finds two factors that strongly seem to influence portfolio choice. The first factor relates to the household balance sheet variables, that is, net worth. The second factor revolves around household personal characteristics such as the overall level of risk aversion. Scherer concludes by recommending future development of the robo-advisory process to be centered around these two key factors (Scherer 2017).

Which direction will future research into robo-advisory take? Glaser et al. (2019) are setting out to conduct a laboratory experiment with 200 individuals to gain insight into the optimal design of robo-advisors for the purposes of determining appropriate ways to present risk to investors. It is the aim to generate an understanding of how different types of individuals understand the concept of risk. Individuals will be tested regarding their grasp of statistical properties and their capability to regulate their own emotions. To allow conclusions, Glaser et al. (2019) hope to measure the confidence, decision times ,and heart rates of participants and to measure changes depending on how and how much information is presented.

This volume provides research that connects to many of the strands of inquiry outlined in the review above and develops additional specialized fields that go to the heart of what characterizes robo-advisory. Starting with a review of the performance of robo-advisory solutions, insights are then provided into goal-based investing and rule-based investing in the context of automated digital services. How can robo-advisory devise and execute appropriate strategies in a context of goals and rules? Contributions are also made to the question of how robo-advisors can enable customers' ownership of the decision-making process when structuring portfolios. This question is linked to the following contribution shedding light onto the topic of portfolio recommendations. Which asset classes lend themselves to be included in robo-advisory recommendations how granular should asset classes be sub-segmented ,and which are appropriate rebalancing intervals? Finally, forays are made into potential future regulation of the advisory space, sales strategies along the human versus machine continuum, as well as bleeding-edge aspects such as artificial intelligence, big data, and social networks.

References

Andrew Arwas and Katie Soleil. 2016. "Robo-Advice 2.0: The Next Generation". *Journal of Financial Transformation* 43: 30–36.

Teodoro Cocca. 2016. *LGT Private Banking Report 2016*. Tech. rep. LGT Group Foundation.

Teodoro Cocca. 2016. "Potential and Limitations of Virtual Advice in Wealth Management". *Journal of Financial Transformation* 44: 45–57.

Richard D'Aveni. 2002. "The Empire Strikes Back. Counterrevolutionary Strategies for Industry Leaders". *Harvard Business Review* 80(11): 66–74.

Gregor Dorfleitner Lars Hornuf, Matthias Schmitt, and Martina Weber. 2017. *FinTech in Germany*. Springer.

Teresa Epperson, Bob Hedges, Uday Singh, and Monica Gabel. 2015. *Hype vs. Reality: The Coming Waves of "Robo" Adoption*. Research rep. AT Kearney. https://www.atkearney.de/financial-services/article?/a/hype-vs-reality-the-coming-waves-of-robo-advisors.

European Parliament and the Council of the European Union. 2015. *Directive (EU) 2015/2366 of the European Parliament and of the Council of 25 November 2015*. Official Journal of the European Union. https://eur-lex.europa.eu/legal-content/EN/TXT/PDF/?uri=CELEX:32015L2366&from=EN.

Michael Faloon and Bernd Scherer. 2017. "Individualization of Robo-Advice". *The Journal of Wealth Management* 20(1): 30–36. https://doi.org/10.3905/jwm.2017.20.1.030.

Melanie L. Fein. 2015. "Robo-Advisors: A Closer Look". *SSRN Electronic Journal*. https://doi.org/10.2139/ssrn.2658701.

Roberto Ferrari. 2016. "FinTech Impact on Retail Banking—From a Universal Bank ing Model to Banking Verticalization". In *The FinTech Book: The Financial Technology Handbook for Investors, Entrepreneurs and Visionaries*. ed. by Susanne Chishti and Janos Barberis. Wiley Online Library, pp. 248–252. https://doi.org/10.1002/9781119218906.ch65.

Florian Glaser, Zwetelina Iliewa, Dominik Jung, and Martin Weber. 2019. "Towards Designing Robo-advisors for Unexperienced Investors with Experience Sampling of Time-Series Data". In *Information Systems and Neuroscience*, pp. 133–138. Springer.

Natalia A. Gold and Steven R. Kursh. 2017. "Counterrevolutionaries in the Financial Services Industry: Teaching Disruption—A Case Study of RoboAdvisors and Incumbent Responses". *Business Education Innovation Journal* 9(1): 139–146.

Dominik Jung, Verena Dorner, Florian Glaser, and Stefan Morana. 2018. "Robo-Advisory: Digitalization and Automation of Financial Advisory". *Business & Information Systems Engineering* 60(1): 81–86.

Dominik Jung, Verena Dorner, Christof Weinhardt, and Hakan Pusmaz. 2018. "Designing a Robo-Advisor for Risk-Averse, Low-Budget Consumers". *Electronic Markets* 28(3): 367–380. https://doi.org/10.1007/s12525-017-0279-9.

Dominik Jung, Edgar Erdfelder and Florian Glaser. 2018. "Nudged to Win–Designing Robo-Advisory to Overcome Decision Inertia". In *Proceedings of the 26th European Conference on Information Systems (ECIS 2018)*, pp. 23–28. https://aisel.aisnet.org/ecis2018%5C_rip/19.

Mehmet Kilic, Peter Heinrich, and Gerhard Schwabe. 2015. "Coercing into Completeness in Financial Advisory Service Encounters". In *Proceedings of the 18th ACM Conference on Computer Supported Cooperative Work & Social Computing*, pp. 1324–1335. ACM. https://doi.org/10.1145/2675133.2675289.

Veith Koerner and Hans-Dieter Zimmermann. 2000. "Management of Customer Relationship in Business Media—The Case of the Financial Industry". In *HICSS '00 Proceedings of the 33rd Hawaii International Conference on System Sciences*, Vol. 6, p. 6045. IEEE. https://doi.org/1.1109/HICSS.2000.926866.

Alexander Maedche, Stefan Morana, Silvia Schacht, Dirk Werth, and Julian Krumeich. 2016. "Advanced User Assistance Systems". *Business & Information Systems Engineering: The International Journal of WIRTSCHAFTSINFOR MATIK* 58(5): 367–370. https://doi.org/10.1007/s12599-016-0444-2.

A. Meola. 2017. *Robo Investing Reviews: Performance & Fees Comparison*. Tech. rep. Business Insider.

Till Moewes, Thomas Puschmann, and Rainer Alt. 2011. "Service-Based Integration of IT-innovations in Customer-Bank-Interaction". In *Wirtshaftinformatik Proceedings 2011*. Paper 102, ed. Bernstein, A., Schwabe, G. http://aisel.aisnet.org/wi2011/102.

Rebecca Nueesch, Thomas Puschmann, and Rainer Alt. 2014. "Realizing Value From Tablet-Supported Customer Advisory: Cases From the Banking Industry". In *BLED 2014 Proceedings*, p. 34. https://aisel.aisnet.org/bled2014/34.

Philipp Nussbaumer Inu Matter Gian Reto a Porta, and Gerhard Schwabe. 2012. "Designing for Cost Transparency in Investment Advisory Service Encounters". *Business & Information Systems Engineering: The International Journal of WIRTSCHAFTINFORMATIK* 4(6): 347–361. https://doi.org/10.1007/s12599-012-0237-1.

Thomas Puschmann. 2017. "Fintech". *Business & Information Systems Engineering: The International Journal of WIRTSCHAFTSINFORMATIK* 59(1): 69–76. https://doi.org/10.1007/s12599-017-0464-6.

Christian Ruf, Andrea Back, Richard Bergmann, and Michael Schlegel. 2015. "Elicitation of Requirements for the Design of Mobile Financial Advisory Services–Instantiation and Validation of the Requirement Data Model with a Multi-Method Approach". In *2015 48th Hawaii International Conference on System Sciences*, pp. 1169–1178. IEEE. https://doi.org/10.1109/HICSS.2015.142.

Bernd Scherer. 2017. "Algorithmic Portfolio Choice: Lessons from Panel Survey Data". *Financial Markets and Portfolio Management* 31(1): 49–67. https://doi.org/10.1007/s11408-016-0282-8.

Paolo Sironi. 2016. *FinTech Innovation: From Robo-Advisors to Goal Based Investing and Gamification*. John Wiley & Sons.

Michael Tertilt and Peter Scholz. 2018. "To Advise, or Not to Advise—How Robo-Advisors Evaluate the Risk Preferences of Private Investors". *The Journal of Wealth Management* 21(2): 70–84.

Ann S. Woodyard and John E. Grable. 2018. "Insights into the Users of Robo-Advisory Firms." *Journal of Financial Service Professionals* 72(5): 56–66.

PART II

Implementation of Robo-Advisory

CHAPTER 3

Risk Preferences of Investors

Monika Mueller, Paul Resnik, and Craig Saunders

Contents

M. Mueller (✉)
Founder, FCM Finanz Coaching, Wiesbaden, Germany
e-mail: monika.mueller@fcm-coaching.de

P. Resnik
Co-Founder, FinaMetrica Risk Profiler, Caulfield North, VIC, Australia
e-mail: paul@paulresnik.com

C. Saunders
Editor, Financial Suitability Forum, Caulfield North, VIC, Australia
e-mail: craig@cbkcommunications.com.au

© The Author(s) 2021

P. Scholz (ed.), *Robo-Advisory*, Palgrave Studies in Financial Services Technology, https://doi.org/10.1007/978-3-030-40818-3_3

3.1 Introduction

"Risk preferences", as a term, are an apparition. They permeate many discussions about financial advice all around the world. Yet, as you approach them, they turn to mist. There is no consensus on a single definition of risk preferences and therefore they do not have a real meaning—each author shapes them to fit their own purpose. This problem of having a poor definition arises because "risk preferences" have emerged as an explanation of why something is not, rather than to define something that is. The problem that risk preferences need to solve is the failure of some retail investors to act rationally when it comes to risk, as the dominant "expected utility" theory said they should (Cohn et al. 1975). Sometimes, they would be more risk-averse than the expected utility model would predict. Clearly, there is something else at play. That something else has evolved to become known as risk preferences. But the term has become a catch-all and melting pot of various explanations, motivations, and observations that might help to explain the investor's aversion to maximize returns by accepting risk. Today, the melting pot still contains the same basic ingredients, which include both psychological and financial elements. But its name has changed. In commercial settings, the term *risk profile* has largely superseded and replaced risk preferences in the discussion; it also accommodates factors that might influence and/or constrain the investor's willingness to accept risk (Klement 2015). However, while the name has changed, the problems of having a poor definition and inconsistent usage persist, with risk profile and risk profiling commonly meaning different things to different people (Moore 2017). Even regulators, who require risk profiling to be conducted, are inconsistent in their rules about what should be included and how profiling should be performed. For example, even though Australian regulations require risk profiling, the Financial Ombudsman service recognizes that experienced advisors may be able to secure a client's informed consent even without using a risk-profiling tool (Financial Ombudsman Service 2011). We, therefore, think it useful to provide our definitions of the key risk terms that will be used in our discussions of risk profiling and align with those presented in a review of global risk profiling best practices (Brayman et al. 2015):

Risk Tolerance
Risk tolerance is the foundation stone of a risk profile. Risk tolerance describes the willingness or aversion of the client to take on investment risk. It is a constant psychological trait that can be measured using psychometric testing.

Risk Capacity
Risk capacity is an accounting measurement of the investor's financial ability to endure investment losses. The question of what is "endurable" is usually measured against whether losses would result in a requirement for any adverse lifestyle changes or abandonment of goals.

Risk Required
Risk required, also referred to as "risk needed", is a mathematical calculation of the level of risk necessary to produce the investment returns necessary to achieve goals. A person with large goals but small assets may require larger amounts of risk than another with the same goals who is starting with more substantial assets.

Risk Composure
This is an estimation of the likelihood that a client perceiving a crisis might act differently than their rational selves, leading them to liquidate investment positions and crystallize losses.

Risk Profile
The risk profile aggregates the preceding factors, along with other secondary factors, to determine the optimal level of risk that maximizes the client's potential to reach their goals while remaining consistent with the level of risk they can financially afford to take and are psychologically willing to accept.

3.2 Risk Profiling Explained

Broadly, a person's risk profile should measure and record their emotional and financial capacity to take investment risks. Preparing an accurate profile is a multidimensional process by combing both objective and subjective elements. An investor's risk profile plays a critical role in making investment decisions and directly influences asset allocations (Droms and Strauss 2003). It is, therefore, of primary importance in making investment recommendations and giving financial advice. A risk-tolerance measurement on its own does not tell the full story of a person's risk preferences which might better be called their "risk profile". There are three critical core components that intersect to form a person's risk profile:

1. Risk tolerance
2. Risk capacity
3. Risk required

A person's risk profile emerges from the interaction between these core aspects, together with other considerations. These include the investment time horizon, which can impact risk capacity, or the knowledge of financial products. The knowledge factor is controversial as it is unconnected with risk tolerance or capacity; it is difficult to quantify and can be readily ameliorated by instruction and explanation.

3.2.1 Financial Risk Tolerance

Financial risk tolerance is a measure of a person's ability to remain invested in assets despite sharp changes in value. Remaining invested throughout volatility is important as it keeps exposure to markets and avoids materializing losses. Financial risk tolerance is a psychological trait that is stable, constant, and predictable; it is largely beyond our own direct control. People do not generally change between high-risk and low-risk tolerance, just as today's extrovert is unlikely to become tomorrow's introvert or vice versa. Risk tolerance tends to be constant over time, except when dramatic events lead to a change in the core personality (Van de Venter et al. 2012). A European study of risk preference in 1507 adults subjected them to 39 different risk-profiling test instruments, with a retest of a sample of 109 participants six months later, thereby confirming the stability of risk tolerance when revealed by psychometric testing (Frey et al. 2017).

A risk-tolerance score is an important starting point for mapping risk tolerance to investment decisions. But ultimately, a person may quite legitimately make investments that depart from those suggested by their risk tolerance score. For example, someone with a lower risk-tolerance score might decide to take on a higher level of risk than they would normally be comfortable with because not doing so would mean they cannot possibly reach their goals. Conversely, a person with high-risk tolerance might hold low-risk investments if they already have enough money to meet their needs and goals. Meanwhile, a person, regardless of risk tolerance, might not be able to afford to bear any financial loss, so they restrict themselves to low-risk, capital-guaranteed investments. There are several approaches to determine a risk score, with some of the methods subject to highly critical concerns about their validity and reliability. These concerns will be explored when we discuss the criteria for selecting a risk-tolerance assessment methodology.

3.2.2 Risk Capacity

Risk capacity is the amount of loss a person can incur without jeopardizing the achievement of important financial goals. It is an accounting measurement that is uncorrelated with risk tolerance—a person with a high level of risk capacity may have a low-risk tolerance and vice versa. Factors which influence risk capacity include the following:

- Age and work status, which can directly impact investment time frames
- Current assets
- Current income
- Future income requirements
- Future inflows and outflows of cash
- Future resources such as pensions or inheritances
- Future liabilities such as taxation, aged care, and family care

There is a lot of interactivity and discretion within these variables. The investor has a lot of direct control over their future requirements and situation. For example, they may be able to adjust down their spending expectations or increase income by choosing to delay their retirement for a number of years.

3.2.3 *Risk Required*

Risk required is the amount of investment risk needed to generate the returns required to meet the investor's objectives, given their starting position. It is an analytical calculation. A simple example demonstrates the principle more easily than a lengthy explanation. Assume a person wants to have $ 110 a year from now, but today they only have $ 100. So, they require a risk/return of 10% to achieve their objective. But if their starting point is only $ 95, the risk/return required increases to 15.8%. Risk required is dynamic and controllable as different variables can be manipulated by the investor. Returning to our simple example, this investor could reduce the risk required by lowering their return expectations from $ 110 to $ 108, or they can simply extend their investment horizon from one year to two.

3.2.4 *Risk Profiling Is Part of Knowing Your Client*

Risk profiling sits within the "know-your-client" (KYC) requirements for financial advice and product sales that are well established in the United Kingdom (Financial Conduct Authority 2006), Europe (European Securities and Markets Authority 2018), the United States (Financial Industry Regulatory Authority 2011), Canada (Fung et al. 2013), and Australia (Australian Securities & Investments Commission 2017). The KYC rules are universally based on the premise that providers who know certain details about a client will be better able to provide advice and products that are suitable for that client, given their unique personality, circumstances and goals. However, the regulations are principle-based and lacking in detail. They speak of the outcomes or standard that should be met but generally say very little about the methodologies or processes that might lead to those goals. Risk tolerance is a key example. All jurisdictions require it to be assessed, yet none explains how to measure it, how to interpret the results, or how it should be "taken into account" while making a recommendation.

3.3 Choosing a Valid and Reliable Risk-Tolerance Test

Firms using risk-tolerance tests must ensure that the testing instrument they are using is fit for the purpose—it should be both valid and reliable. To be valid, a test must be true to the label, which means it must measure what it claims to measure—in this case, financial risk tolerance. To be reliable, the test results must be largely consistent regardless of how, where, or when the test is administered (Grable and Lytton 1999). Many of the risk-tolerance tests in use today fail to meet either of these standards. A 2015 review of global best practices found that only 16.7% of the questionnaires reviewed would be considered fit for the purpose. Around 80% of the tests had been developed in-house or were left to individual advisors' discretion. Regulators typically require that advice-givers conduct initial and ongoing reviews of the tools used in the advice process (Financial Industry Regulatory Authority 2016). That review should include questions covering the following:

- Whether methodologies and assumptions are well suited to the task
- If methodologies are tested by independent third parties
- What data points will be used and how
- How outputs will be tested to ensure they conform with expectations
- Whether the models used remain appropriate as market conditions change
- How the tool identifies and deals with exceptions and inconsistencies

Very few in-house risk-tolerance test instruments have been developed according to such strict criteria. The 2015 global review (Brayman et al. 2015) found that less than 10% of firms knew if their test had been validated by a statistical check of reliability. Less than 20% had engaged any outside expertise to assist in developing their test instruments.

Many in-house questionnaires fail to meet the criteria for validity as they contain questions that pertain to something else other than risk tolerance, such as risk capacity. Reliability is also a problem as many tests do not ask enough questions to produce meaningful results (Roszkowski et al. 2005).

3.3.1 Psychometrics Versus Gambles-Based Testing

Two dominant methodologies are used by commercial providers of risk-tolerance tests:

1. Psychometric testing of risk tolerance as a personality trait, drawing on an extensive body of psychological research around psychometrics (Rust and Golombok 2009);

2. Gambles-based questionnaires about wins versus losses, drawn from the prospect theory (Kahneman and Tversky 1979).

The divide is sharp—providers' products tend to be rooted in either psychometric or gambles-based methodologies, but generally not both.

To express a belief in one methodology is essentially to express disbelief in the other. This does, in fact, make intuitive sense. Supporters of psychometrics argue that the task tests a stable psychological trait, while gambles-based supporters argue that the task determines preferences from a series of choices. They are very different approaches. The debate over which methodology is effective and/or superior is just as sharp with somewhat of a "tribal" flavor. There is little overlap between the two, and the debate tends toward polarization. This debate has been going on for many decades. A series of due diligence questions about a risk-tolerance test has been developed by a team, including psychometrics specialists, from the London School of Economics (Resnik 2016). It puts forward questions to be asked about the face validity of the test; expertise of the people who created it; independent expert scrutiny and testing; reliability of the assumptions relating to constructs underpinning the test; and the usability of the test for the client (Erskine 2016).

3.3.2 The Science of Prospect Theory

Kahneman's and Tversky's Nobel prize-winning prospect theory is a groundbreaking work that has amassed many hundreds of thousands of academic citations. Prospect theory describes how people choose between probabilistic alternatives involving risk where the probabilities of outcomes are uncertain. It says that people use heuristics or mental shortcuts to evaluate gains and losses, which they value more highly than the final outcome. The theory was developed partly to address the failure of the expected utility theory to accurately describe consumer behavior. It is seminal in the field of behavioral economics and widely respected. But the use of prospect theory in the assessment of financial risk tolerance is controversial, with concerns about both validity and reliability. A key validity concern is the integrity of data gathered through gambles-related questions, given the difficulty that many people face in providing a meaningful answer to a question involving complex mathematical topics such as probability.

Much of the population lack the capacity to complete a calculation of this type, while others will avoid the question because of the "math anxiety" it provokes (Ashcraft 2002). In the absence of a meaningful response, the investor may resort to guessing an answer, thereby undermining the efficacy of the test (Baloğlu 2004). Research has also shown that even in simple lotteries people tend to rely on heuristics, rather than making the necessary calculations, to understand the risk and reward relationship (Cokely and Kelley 2009). The hypothetical nature of the gamble is also a concern for the validity. Research has shown that people can behave very differently when stakes are real and meaningful compared to stakes that are hypothetical and immaterial (Barreda-Tarrazona et al. 2011)—this raises doubts over the answers to gambles-related questions (M. Levy and H. Levy 2002). Reliability is also an issue for gambles-based risk-tolerance tests, with anecdotal evidence of variances in results upon retest.

Meanwhile, some in the financial services sector feel some kind of basic discomfort in forming an association between gambling and investing in their clients' minds. Many have fought hard to distance financial advice from the "stock-market is a casino" idiom, and they have been reluctant to introduce "gambles" into a disciplined investment process.

3.3.3 *The Science of Psychometrics*

Psychometric testing of risk tolerance has been subject to extensive independent reviews and testing; it has been shown to be a valid and reliable methodology (Hallahan et al. 2003). Psychometrics is an amalgam of psychology and statistics that can quantify and assess psychological traits and constructs. While its heritage has some controversy, it is today a well-established and accepted science with validation tools to determine the technical quality of psychological assessment tools like questionnaires.

However, many questionnaire tools that claim to be psychometric are not so—indeed, they may not have any scientific basis at all. For example, the Myers–Briggs Type Indicator (MBTI) is a self-report questionnaire which is the most widely used personality test in the world, with around two million people a year undertaking the questionnaire. It is commonly used during recruitment processes to identify "personality types" that explain how people perceive as well as respond to the world around them. Unfortunately, this test consistently fails the key scientific requirements of psychometrics as it displays poor validity and reliability. Poor validity manifests as not measuring what the test proposes to measure, not having any predictive power, or not having any items that can be generalized. Poor reliability means that the test can produce different results for a person taking the test at different times and not being comprehensive (Grant 2013). These problems can be traced to the development for the MBTI test, which did not use any scientific processes. This test was developed by Katharine Cook Briggs and her daughter, Isabel Briggs Myers, who based their work on the theories of the psychiatrist Carl Jung, whose work was observational and deductive. Katharine began developing her personality-types framework in 1917 based on her own reading of the literature as an introspection. However, her work then languished for more than 20 years. Katharine and her daughter did not create a test until 1944, when it found traction as a tool to help place women into appropriate industrial jobs in America during World War II. True psychometric testing follows a scientific development process, which includes extensive testing of the test instrument (the questions) and the "proving" of results for validity and reliability.

Psychometric testing of financial risk tolerance emerged in 1999 with the development and launch of FinaMetrica (T. G. Davey 1999)—the first custom-built psychometric test for risk tolerance that was launched in Australia and is now used globally in more than 20 countries. Other providers of psychometric tests have since entered the global market. The FinaMetrica test was developed using a scientific process. More than 100 questions were tested in rigorous academic environments and processes, with most being discarded as they were not dependable in producing valid and reliable results. The final test was reduced to 25 questions where confidence about validity and reliability was very high. The FinaMetrica psychometric test of financial risk tolerance produces a "score", which places the investor with a continuum of range from no risk at all at one extreme to 100% risk appetite at the other. But in reality, almost no one ever appears at those extremes. Most people are fairly similar and tend to cluster around the middle, with only a few outliers showing up near the very conservative or very risky ends. This is a "bell curve" or normal distribution pattern (Katsikatsou and Erskine 2018). The curve shows that 86% of people fall within one standard deviation of the "middle" (Fig. 3.1). But while the differences between people may appear small in statistical terms, they can nonetheless become very significant in real life—a person with a score of 37 will be very different to one with a score of 57, even though both are equally close to the middle. Almost 1.5 million

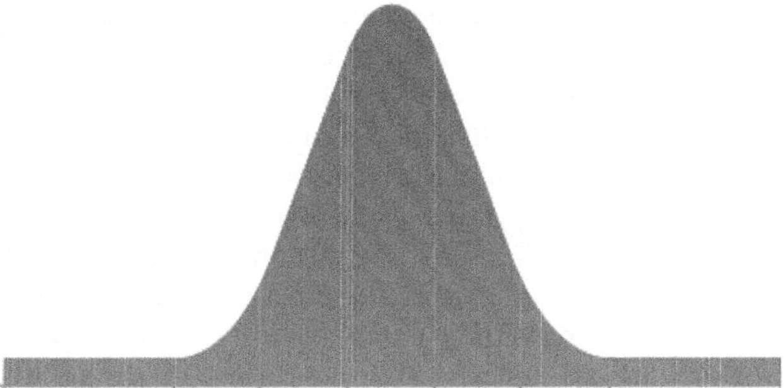

Risk Group	1	2	3	4	5	6	7
Score Range	Less than 25	25-34	35-44	45-54	55-64	65-74	75 or more
No in group	1%	6%	24%	38%	24%	6%	1%

Fig. 3.1 Financial risk tolerance is a stable psychological trait which follows a normal distribution pattern, where most people cluster around the middle peak (the "norm") with reducing numbers of outliers at each end (the very risk averse and extreme risk seekers)

people have had their risk tolerance measured using the FinaMetrica test instrument. This data, after being deidentified, has been made available for academic analysis, testing, and review. It has been verified for validity and reliability.

3.4 Risk Profiling Is Harder for Robo-Advisors

Robo-advisors must comply with the same rules and regulations as other financial advice systems, with all major regulators having made statements that the rules governing financial advice are the same regardless of how it is delivered. This means human-driven advice systems, cyborg advice systems (where a human is augmented by, or augments, technology), and robo-advisors must all meet the same standards for risk profiling. Everyone faces the same initial hurdle: selecting a fit-for-purpose risk-tolerance assessment methodology and instrument on which to build the profiling process. But the risk-profiling job is harder for robo-advisors, and some have failed to meet the required standards (Smith 2016).

Challenge 1: Understanding and Empathy

The first challenge the robo faces is the "warm body effect" (Fisch et al. 2018). The robo suffers from the lack of insight and flexibility that a human can bring. A human advisor brings their presence, expert judgment, and emotional intelligence to an encounter with a client. They "read" the client's speech, actions, and body language and inevitably make many minor adjustments in how the meeting proceeds. Based on what they see and hear, they might pause to spend extra time on helping clients with a matter of financial literacy, consider adjusting goals or attitudes, or address apparent inconsistencies in the facts before them. A robo-advisor, however, is the opposite in virtually every respect. A human knows a client as a person. But a robo-advisor only knows a client as a set of data points. For a robo-system, the only inputs from the client are predefined and narrow. Often, they must select one of four prewritten answers as their "best fit" answer to a question. The robo cannot "see" or otherwise intuit how the client looks, sounds, or seems as they provide the answer. The only input is that single "best-fit" data point. So, the robo-advisor cannot respond dynamically as a human might. A human might sense that help is needed and accordingly offer it. But a robo-advisor will only offer help when the user specifically asks for it or a pre-coded alarm is triggered by the answer to a particular question. The immediate problem is not how to make robo-advisors more human because research has shown that this can actually diminish the client's confidence in the advice given by the robo (Hodge et al. 2018). Intriguingly, the research found people are more prepared to follow advice from a robo when it exhibits fewer human characteristics. Rather, the immediate problem is how to equip robo-advisors to understand and deal with the subtleties and nuances that a client might experience during an automated encounter. A research study (Glaser et al. 2019) was announced in 2018, focusing on robo-advisors that can "speak" to investors in plain language in an intuitive way and "listen" to the investor by monitoring their emotional reactions. The answers will lie in creating "smarter" robos with sophisticated algorithms that can create a "view" of the investor as a person whose sum is greater than the parts of their financial data points.

Challenge 2: Building More Comprehensive Algorithms

Building more comprehensive algorithms is hard. It means dealing with known knowns, known unknowns, and unknown unknowns to be considered and accommodated. The mapping can quickly become like a spider web. These expanded algorithms will almost inevitably require extra questions, inputs, or steps. To the designers of online experiences, this is anathema. The tech mantra is to simplify and reduce rather than expanding to add complexity. However, financial advice is an area that defies simplification. Risk-tolerance tests are a good example. A psychometric risk-tolerance test may contain as many as 25 questions. The test is designed to capture the required data, but it will also ensure internal consistency and identify unusual or atypical responses. It requires a large number of questions for completion of all that work. But robo-advice risk-tolerance questionnaires can be as short as just five questions. It is difficult to see that the robo's algorithm can be sufficiently informed by a dataset so small. The problem is not one of intent—the designers of robo-advisors are responding to solid evidence that users of online services want the quickest possible outcomes with the least possible work. "Friction" is to be eliminated. But in a highly regulated, closely monitored industry like financial advice, or aviation, or medicine, there are often no shortcuts. A good process requires a good process, which often requires numerous steps.

Challenge 3: Applying "Professional Judgment"
A human advisor uses their presence, insight, and emotional intelligence to inform the exercise of their professional judgment—often on a case-by-case basis. This approach can introduce problems of inconsistency. But it also creates a greater opportunity to shape the advice to the individual receiving it. How a robo-advisor might approach and apply professional judgment is a key issue to be resolved. Will it be done at all? Should it be done at a "house" level where all advice prioritizes the one variable over others? Or should it be a highly individualized experience, reflecting a sample of only one? Once the approach has been decided, it can be codified into a robo-advisor, which should then produce consistent outcomes based on the algorithm. This can overcome the problems of variability and inconsistency that may arise among human advisors.

3.5 Mapping Risk Profiles to Investments

A risk profile's purpose is to guide investment decisions. However, great care must be taken when mapping a particular risk "score" to appropriate investments. The most immediate danger is that the labels and terminology around risk cannot be trusted or relied upon, as most of them have no defined or agreed meaning. Terms like these can mean almost anything at all—it all depends on how the particular product provider defines it:

- Conservative
- Aggressive
- Defensive
- Balanced
- Low risk
- Medium risk
- High risk

A 70% exposure to growth assets, such as equities, might be called "balanced" by one provider, "aggressive" by another, and "medium risk" by yet another. There is no regulation anywhere in the world governing how or when terms like these should be used.

3.6 Conclusion

Robo-advisors operate in a highly regulated marketplace that is difficult to "disrupt" as robos are obligated to follow the same rules that apply to human advisors. Nonetheless, they are exciting for the opportunities they present to scale financial advice, particularly in sectors that might otherwise be unprofitable to serve. However, robo-advisors face extra challenges. They have fewer dimensions than a human advisor in terms of interaction and dexterity. They must be taught in advance about every possible situation that can be encountered. The underlying algorithms are both the strength and weakness for a robo-advisor. When they are robust, the robo can scale good advice across large numbers of people. But when they are inadequate, the resulting advice can become flawed. Risk tolerance is a particularly good example of a challenge for robo-advisors who face pressure to reduce the number of inputs required from a user.

References

Mark H. Ashcraft. 2002. "Math Anxiety: Personal, Educational and Cognitive Consequences". *Current Directions in Psychological Science* 11(5): 181–185.

Australian Securities & Investments Commission. 2017. *Regulatory Guide 175—Licensing: Financial Product Advisers—Conduct and Disclosure.* https://download.asic.gov.au/media/4698465/rg175-published-10-april-2018.pdf.

Mustafa Baloğlu. 2004. "Statistics Anxiety and Mathematics Anxiety: Some Interesting Differences I". *Educational Research Quarterly* 27(3): 38–48.

Ivan Barreda-Tarrazona, Ainhoa Jaramillo-Gutierrez, Daniel Navarro-Martinez, and Gerardo Sabater-Grande. 2011. "Risk Attitude Elicitation Using a Multi-Lottery Choice Task: Real vs. Hypothetical Incentives". *Spanish Journal of Finance and Accounting* 40: 609–624. https://doi.org/10.108/02102412.2011.10779713.

Shawn Brayman, Michael Finke, Ellen Bessner, John E. Grable, Paul Griffin, and Rebecca Clement. 2015. *Current Practices for Risk Profiling in Canada and Review of Global Best Practices.* Last Accessed: September 13, 2019. https://www.osc.gov.on.ca/documents/en/Investors/iap_20151112_risk-profiling-report.pdf.

Richard A. Cohn, Wilbur G. Lewellen, Ronald C. Lease, and Gary G. Schlar baum. 1975. "Individual Investor Risk Aversion and Investment Portfolio Composition". *The Journal of Finance* 30(2): 605–620. https://doi.org/10.2307/2978738.

Edward T Cokely and Colleen M. Kelley. 2009. "Cognitive Abilities and Superior Decision Making Under Risk: A Protocol Analysis and Process Model Evaluation". *Judgment and Decision Making* 4(1): 20–33.

Thomas G. Davey. 1999. *Automated Assessment of Personal Financial Risk Tolerance.* U.S. Patent US6859788B1. https://patents.google.com/patent/US6859788B1/en.

William G. Droms and Steven N Strauss. 2003. "Assessing Risk Tolerance for Asset Allocation". *Journal of Financial Planning* 16(3): 72–77.

Stuart Erskine. 2016. *Due Diligence Questions About a Risk Tolerance Test.* Last Accessed: September 15, 2019. http://www.riskprofiling.com/WWW_RISKP/media/RiskProfiling/Downloads/Due_Diligence_Checklist.pdf.

European Securities and Markets Authority. 2018. *Guidelines on Certain Aspects of the MiFID II Suitability Requirements.* https://www.esma.europa.eu/sites/default/files/library/esma35-43-1163_guidelines_on_certain_aspects_of_mifid_ii_suitability_requirements_0.pdf.

Financial Conduct Authority. 2006. *FCA Handbook: COB 5.2 Know Your Customer.* Last Accessed: September 13, 2019. https://www.handbook.fca.org.uk/handbook/COB/5/2.html?date=2006-03-01.

Financial Industry Regulatory Authority. 2011. *Regulatory Notice: Know Your Customer and Suitability.* https://www.finra.org/sites/default/files/NoticeDocument/p122778.pdf.

Financial Industry Regulatory Authority. 2016. *Report on Digital Investment Advice.* https://www.finra.org/sites/default/files/digital-investment-advice-report.pdf.

Financial Ombudsman Service. 2011. *Risk Profiling in Financial Advice Disputes.* Last Accessed: September 13, 2019. https://www.fos.org.au/the-circular-6-home/risk-profiling-in-financial-advice-disputes/.

Jill E. Fisch, Marion Laboure, and John A. Turner. 2018. "The Emergence of the Robo-Advisor". In *Wharton University of Pennsylvania Pension Research Council.* Working Paper WP2018-12. https://repository.upenn.edu/prc_papers/10/.

Renato Frey, Andreas Pedroni, Rui Mata, Jorg Rieskamp, and Ralph Hertwig. 2017. "Risk Preference Shares the Psychometric Structure of Major Psychological Traits". *Science Advances* 3(10). https://doi.org/10.1126/sciadv.1701381.

Carlin Fung, Paul Hayward, and Mark Skuce. 2013. *Know-Your-Client and Suitability Obligations.* https://www.osc.gov.on.ca/documents/en/Dealers/ro_20131204_kyc-suitability-%09obligations.pdf.

Florian Glaser, Zwetelina Iliewa, Dominik Jung, and Martin Weber. 2019. "Towards Designing Robo-advisors for Unexperienced Investors with Experience Sampling of Time-Series Data". In *Information Systems and Neuroscience,* pp. 133–138. Springer.

John E. Grable and Ruth H. Lytton. 1999. "Financial Risk Tolerance Revisited: The Development of a Risk Assessment Instrument". *Financial Services Review* 8(3): 163–181. https://doi.org/10.1016/S1057-0810(99)00041-4.

Adam Grant. 2013. *Goodbye to the MBTI, the Fad that Won't Die.* Psychology Today. Last Accessed: September 15, 2019. https://www.psychologytoday.com/au/blog/give-and-take/201309/goodbye-mbti-the-fad-won-t-die.

Terrence A. Hallahan, Robert W. Faff, and Michael D. McKenzie. 2003. "An Empirical Investigation of Personal Financial Risk Tolerance". *Financial Services Review* 13: 57–78.

Frank D. Hodge, Kim I. Mendoza, and Roshan K. Sinha. 2018. "The Effect of Humanizing Robo-Advisors on Investor Judgments". *SSRN Electronic Journal.* https://doi.org/10.2139/ssrn.3158004.

Daniel Kahneman and Amos Tversky. 1979. "Prospect Theory: An Analysis of Decision Under Risk". *Econometrica, Econometric Society* 47(2): 263–291.

Myrsini Katsikatsou and Stuart Erskine. 2018. *FinaMetrica Technical Manual.* https://www.riskprofiling.com/WWW_RISKP/media/RiskProfiling/Downloads/Technic%09al_Manual_2018.pdf.

Joachim Klement. 2015. "Investor Risk Profiling: An Overview". *The CFA Institute Research Foundation*. https://www.cfainstitute.org//media/documents/article/rf-brief/rfbr-v1-n1-1-pdf.ashx.

Moshe Levy and Haim Levy. 2002. "Prospect Theory: Much Ado About Nothing?" *Management Science* 48(10): 1334–1349.

Stephen H. Moore. 2017. "Communication Task Difficulty in Investment Risk Profiling: A Linguistic Perspective". *Financial Planning Research Journal*, 33–52.

Paul Resnik. 2016. *How to Test Your Clients' Risk Tolerance*. FTAdviser.com. Last Accessed: September 13, 2019. https://www.ftadviser.com/investments/2016/9/28/how-to-test-your-clients-risk-tolerance/.

Michael J. Roszkowski, Geoff Davey, and John E. Grable. 2005. "Insights from Psychology and Psychometrics on Measuring Risk Tolerance". *Journal of Financial Planning* 18(4): 66–77.

John Rust and Susan Golombok. 2009. *Modern Psychometrics: The Science of Psychological Assessment*. Taylor & Francis Group. Isbn:9780415442152.

Matthew Smith. 2016. *How Robo-Advice Could Go Bad*. FINSIA InFinance. https://finsia.com/insights/news/news-article/2016/08/11/how-robo-advice-could-go-bad.

Gerhard Van de Venter, David Michayluk, and Geoff Davey. 2012. "A Longitudinal Study of Financial Risk Tolerance". *Journal of Economic Psychology* 33(4): 794–800. https://doi.org/10.1016/j.joep.2012.03.001.

CHAPTER 4

Robo Economicus? The Impact of Behavioral Biases on Robo-Advisory

Peter Scholz, David Grossmann, and Joachim Goldberg

Contents

P. Scholz (✉)
Professor for Banking and Financial Markets, Hamburg School of Business Administration, Hamburg, Germany
e-mail: peter.scholz@hsba.de

D. Grossmann
Head of Global Bank Management, Otto M. Schroeder Bank AG, Hamburg, Germany
e-mail: david.grossmann@schroederbank.de

J. Goldberg
Founder, Goldberg & Goldberg, Frankfurt am Main, Germany
e-mail: joachim@goldberg-goldberg.com

© The Author(s) 2021

P. Scholz (ed.), *Robo-Advisory*, Palgrave Studies in Financial Services Technology, https://doi.org/10.1007/978-3-030-40818-3_4

4.1 Introduction

"Robo-advisors advertise investments without emotional bias, because the algorithm gives rational suggestions. The potential of human failure should be minimized."[1] One important element of the robo-advisory business is indeed to avoid the emotional components like greed, fear, or doubt in the investment process. For example, the largest robo-advisor based on assets under management claims that "[w]hen it comes to investing, your natural reactions can get in the way. It's human nature to overthink, overreact, and, at times, be overwhelmed. With Vanguard Personal Advisor Services, an advisor serves as an emotional circuit breaker so you don't abandon a well-thought-out plan" (Vanguard 2020a).

The idea to evolve machine support for investment advisory is not new: more than 15 years ago, cognitrend, a company specialized in behavioral finance analysis and trading, created an "artificial salesman," who would assist the human advisors based on behavioral economics. The artificial salesman was planned as a web-connected algorithm capable of rational decision-making. Whereas humans tend to be influenced by past successes of failures and are biased by imperfect information processing, machines seem to be unmoved by these pitfalls. But at that time, trust in human financial advisors and skepticism against internet and computer were probably too pronounced; and because of that the artificial salesman failed to catch the fancy of common investors. After the financial crisis, the preferences underwent a major shift and the first robo-advisors appeared in the private and retail segment.

Although robo-advisors keep their focus on the various aspects of investment such as transparency, cost-efficiency, or usability, it is the unerring rationality as well in which they seem to be ahead of their human colleagues. In the area of financial decision-making, there is an almost infinite number of scientific publications available on behavioral finance, which describe the impact of emotions on human decisions. Authors such as Daniel Kahneman ("Thinking, Fast and Slow") or Nassim N. Taleb ("The Black Swan") present numerous examples as to how investors can be trapped by their emotions and limitation in information processing. Infamous shortcomings include anchoring, availability heuristic, disposition effect, gambler's fallacy, and selective perception, to name only a few.

[1] Translated from Grzanna (2018). A similar description can be found on Rixse (2018).

Since machines are created and programmed by humans, it would be interesting to explore if robo-advisors really corresponded to the ideal of Homo economicus, the fully rational decision-maker—a robo economicus, for instance—or if the robots were more similar to the humans who invented them. For this analysis, we took a deeper look into three different behavioral biases: home bias, mental accounting, and overconfidence. Even if we believe that the robo-advisors probably make the investment process less vulnerable to emotional influences, we are interested to see to what degree the improvement evolves. Each section briefly describes the respective bias and then analyzes the status quo as well as the potential exposure of robo-advisory to the bias. Here, we focus primarily on the big players in the US, the UK, and Germany, selected in terms of assets under management, who currently drive the market.

4.2 Home Bias

"Why seek far afield when the good could not be any closer by?"[2] is a famous saying attributed to Johann Wolfgang von Goethe. Ambiguity aversion, that is, to prefer the familiar and abandon the unknown, also seems to hold true on capital markets: as French and Poterba (1991) discovered, investors tend to overweight domestic stocks and neglect the benefit of international diversification. In their sample, the bias seems to be exceptionally evident: for US investors, they found a 0.938 weight in domestic assets, for Japanese investors 0.9811, and for UK investors 0.820. They explain the relatively low weight of UK stocks in UK portfolios compared to the US and Japanese cases with the smaller market value of UK assets and the consequential need to invest in foreign equity. In another paper, Tesar and Werner (1995) analyze home bias in Canada, Germany, Japan, the UK, and the US for stocks and bonds. They emphasize that "there is strong evidence of a home bias in national investment portfolios despite the potential gains from international diversification." Both analyses have been performed with data sets, in which international deregulation and international trading just started to accelerate a few years ago, which might be an explanation for the poor international diversification at that time. But later studies as well, for example Coval and Moskowitz (1999), Ahearne et al. (2004), Fidora et al. (2007), and Lütje

[2] Translation from German to English by dict.leo.org.

and Menkhoff (2007), confirm the persistence of home bias. Furthermore, there is evidence in some of these studies that even professional investment managers are prone to it. This raises the question if overweighting domestic assets is really undesired or if factors like higher transaction costs for foreign assets, double taxation, currency risk, and information asymmetries might explain the rationality of home bias. But many studies show that, in general, poor international diversification seems to reduce a portfolio's performance significantly (e.g., French and Poterba (1991), Tesar and Werner (1995), Lütje and Menkhoff (2007), Seasholes and Zhu (2010)). Lewis (1999) tries to quantify the cost of insufficient international diversification and states that "the costs of holding foreign stocks must be extremely large to dissuade an efficient domestic investor from foreign diversification." Hence, it is more likely that the explanation of home bias traces back to different behavioral biases such as overconfidence and regret aversion, or factors such as patriotism and excess investments in own-companies' shares (Foad 2011).

4.2.1 Does Robo-Advisory Fall for Home Bias?

Clients of robo-advisory services might assume that the portfolio recommendation of the machine is free from home bias. However, since many robo-advisors rely more on investment committees rather than rule-based asset allocation, the home bias could also be evident in portfolios based on robo-advice. As we have detected in our paper (Tertilt and Scholz 2018), robo-advisors seem to absorb certain other culture differences, since US robos typically recommend higher equity quotas than what German robos do. Therefore, we check the recommended model portfolio of the largest U.S., U.K., and German robo-advisors to verify the distribution between domestic and foreign equity. Hence, for our analysis, we solely consider model portfolios with a close to 100% investment in equity for maximal visibility of the effect.

Considering the largest robo-advisory in the world, Vanguard Personal Advisor Services, the analysis starts with a bit difficult since it is necessary to open an account before a model portfolio is displayed. Furthermore, Vanguard's robo is more a hybrid advice model since the human advisor as well as a human investment committee play a crucial role in the investment process. If we take the Vanguard ETF strategic model portfolio as the next best guess,[3] we find that in an 100% equity portfolio the share of domestic stocks is 58.8% compared to 39.2% of foreign stocks. If we compare these numbers to the other large US robo-advisors, then we see that the figures seem to be in line with the model portfolios of the other robos. Charles Schwab Intelligent Portfolios has a domestic equity quota of 53.2%. Furthermore, Schwab's robo-advisor, as well as Vanguard's, relies on human investment committees. The two robo-advisors with the longest track record, Betterment and Wealthfront, however, use rule-based algorithms for their portfolio allocation. Still, their equity quotas are not so different from those of Vanguard and Schwab: they allocate 48.2% and 46.9% in domestic stocks, respectively. Finally, Personal Capital, which is a digital wealth management service, also relies on an investment committee and, additionally, on a team of experts, including Nobel laureate Harry Markowitz, who is the "father of modern portfolio theory," and Shlomo Benartzi, a distinguished expert on behavioral finance. Interestingly, their quota on domestic stocks is the highest amongst all robo-advisors: they invest 67% in domestic stocks, which is even larger than the North American equity share in the MSCI World Index.

It is doubtless a tricky thing to clearly determine home bias. But if we take the MSCI World Index as reference, which is a market cap-weighted index, then they currently allocate around 56.7% to North American stocks. Hence, Personal Capital would have a clear home bias in their portfolio. Vanguard and Schwab are close to the MSCI benchmark, and even Betterment and Wealthfront are not far away from this reference level. If we consider the share of global gross domestic product (GDP) as a benchmark, which attributes approximately 25.9% to North America, all US robos would show a distinct home bias. Further, it is interesting to note that nearly all tested European robo-advisors show lower North American equity quotas than those of their US peers, which also strengthens the suspicion of a home bias of US robo-advisors.

If we analyze the UK robo-advisors, we find that, especially, Nutmeg seems to have a rather high domestic stock quota: they hold close to 25% in UK stocks. Compared to a weight of 5.7% in the MSCI World Index or 3.3% in share of global GDP, this is comparatively high and seems to qualify for a home bias. Moneyfarm has a lower domestic stock

[3] Based on the assumption that the investment committee makes similar decisions in their ETF portfolio and the robo portfolio.

allocation with around 11.6%, but still this seems to be high in comparison to the significance of British markets or economy in the world. For the German robo-advisors, we find a similar picture. Since they are part of the Eurozone, it makes sense to widen the comparison to European stocks instead. With respect to the MSCI World Index (14.4%), Scalable (24.9%), Weltinvest (19.7%), and Liqid (22.7%) overweight European stocks. If we take the share of the global GDP as benchmark (21.9%), the picture changes and the domestic stock of German robo-advisors seems to be in line. Hence, for German robo-advisors, the picture is not clear regarding home bias.

4.3 Mental Accounting

"29,086 measures barley 37 months Kushim" dates back around 5000 years and is one of the oldest records from Mesopotamia that our forefathers left for us (Harari 2015). So, the oldest messages from our past are not philosophical or religious, and they do not contain some kind of deeper wisdom; as it seems, they are simply accounting statements (Harari 2015). Hence, the compulsion of keeping track of one's own belongings seems to be deep-rooted in human culture. Today, in our highly organized world, accounting is self-evident. As Richard Thaler (1985, p. 199) states: "All organizations, from General Motors down to single person households, have explicit and/or implicit accounting systems." From a certain perspective, this is quite logical: Accounting allows organizing even complex capital flows and was one important factor behind the emergence of North Italian economy in the medieval period. But, as Thaler (1985, p. 199) also notes, "[t]he accounting systems often influence decisions in unexpected ways." Based on Kahneman and Tversky (1979) Prospect Theory, and his considerations of human decisions involving, for example, sunk costs or opportunity costs (Thaler 1980), he found a human behavioral trait that he named mental accounting (Thaler 1985). The first example he presented in his 1985 paper describes the bias (p.199):

> Mr. and Mrs. L and Mr. and Mrs. H went on a fishing trip in the northwest and caught some salmon. They packed the fish and sent it home on an airline, but the fish were lost in transit. They received $ 300 from the airline. The couples take the money, go out to dinner and spend $ 225. They had never spent that much at a restaurant before.

Thaler (1985) explains:

> Example 1 violates the principle of fungibility. Money is not supposed to have labels attached to it. Yet the couples behaved the way they did because the $ 300 was put into both "wind fall gain" and "food" accounts. The extravagant dinner would not have occurred had each couple received a yearly salary increase of $ 150, even though that would have been worth more in present value terms.

The similar effect is described in Tversky and Kahneman (1981). They present the result from an experiment, in which the test persons had to answer two questions. Question 1 is:

> Imagine that you have decided to see a play where admission is $ 10 per ticket. As you enter the theater you discover that you have lost a $ 10 bill. Would you still pay $ 10 for a ticket for the play?

Question 2 asks:

> Imagine that you have decided to see a play and paid the admission price of $ 10 per ticket. As you enter the theater you discover that you have lost the ticket. The seat was not marked and the ticket cannot be recovered. Would you pay $ 10 for another ticket?

Although both questions relate to the same economic situation, the test persons react differently: whereas in the first scenario almost everyone (88%) tends to buy a ticket again, in the second situation only 46% decide to do so. They assign this bias in behavior to the effect of "psychological accounting," which is just another expression for mental accounting. So, in a nutshell, people have the tendency to create imaginary accounts when making decisions. Since people are affected more by losses than by profits, according to the Prospect Theory, the transfer of money from one to another imaginary account may severely bias decision-making in the same economic framework.

4.3.1 *Does Robo-Advisory Care About Mental Accounts?*

It is a valid question if mental accounting is always an unfavorable heuristic or not. As Shefrin (2000) points out, there are two different types of investors: On the one side, there are "[m]ean-variance investors [who] care only about the expected returns and variance of the overall portfolio [and...] have consistent attitudes towards risk." On the other side, there are "[b]ehavioral investors [who] build portfolios as pyramids of assets, layer by layer, where layers are associated with particular goals and particular attitudes towards risk." Many investment advisors follow the mean-variance investor's approach and try to measure the total risk tolerance of an investor, which will be applied to the total assets. However, as Pan and Statman (2012) claim, "each investor has a multitude of risk tolerances. Probing for one global risk tolerance misses that multitude." By and large, investors may pursue different goals with their investments, such as pension, reserves, education for their children, or to make a dream come true (c.f. also Pan and Statman (2012)). It seems to be a fair assumption that "investors consider their portfolios as collections of mental accounts, each devoted to a goal" (Pan and Statman 2012). Depending on the goal's time frame, different risk capacities are possible: The longer the investment period, the higher the risk class of the assets can be selected. The German Institute for Equity publishes the "Return-Triangle" in which for different investment periods the resulting returns are displayed (Deutsches Aktieninstitut 2020). They show that the longer the investment period, the lower is the risk of losing money with stock investments. From this perspective, it makes sense to distinguish between different investment goals by different risk tolerances. For example, while the reserves should show lower risk, long-term pension plans can bear more risky assets. As a consequence, robo-advisory could prefer goal-based investing as proposed by Chhabra (2005), Das et al. (2010), and Brunel (2011), among others.

Interestingly, the large US robo-advisors indeed pursue strategies for goal-based investing. Vanguard Personal Advisor Services proposes different goals as an integral part of its advice and differentiates between its reports based on various investment goals (Vanguard 2020c). It even has a website dedicated to different investment goals and appropriate consideration in individual investment plans (Vanguard 2020b). Also, the US competitors have a clear focus on goal-based investing: Schwab Intelligent Portfolios provides a systematic goal-tracking tool (Charles Schwab & Co. Inc. 2020). The goal-tracker aims to "monitor whether your goal is 'on target' towards achieving a savings or income goal." Schwab's robo-advisor as well provides a website, which contains support regarding investment goals (Charles Schwab & Co. Inc. 2016). They explicitly introduced a bucket system with different time horizons: from less than two years (bucket 1) to three to ten years (bucket 2), and more than ten years (bucket 3). Each of these buckets pursues different strategies and seems to allow for different risk tolerance. Betterment also follows the idea of goal-based investing. Like their peers, it has a detailed website introducing investors to the concept. Betterment clearly states that "[w]ithin your Betterment account, every investment goal you set has a target amount and target date(s) for which you desire to meet you goal [... and] [e]ach of these investment goals requires a different strategy—that is the quintessence of smarter investing" (Egan 2019). Wealthfront aims to go even a step beyond its peers and plans to introduce "self-driving money." As Wealthfront CEO Andy Rachleff stated: "Our vision is to deliver a service where you direct deposit your paycheck with us. We automatically pay your bills. We automatically top off your emergency fund, and then route money to whatever account is the most ideal for your particular goals, whether they're at Wealthfront or elsewhere" (Thiagarajan 2018).

In contrast to their US peers, the European robo-advisors seem to focus less on the goal-based approach. Nutmeg, for example, asks for financial goals in its questionnaire: "When you sign up, we first ask you about your goals and risk preference. With that information, we help you choose a portfolio that's right for you [...]" (Nutmeg 2020). But it does not become clear on its website, how it exactly uses the information and if multiple goals are allowed. Scalable claims to use goal-based investing (Scalable Capital 2018), but on its website we could not find much information about how this concept was applied. Scalable asks in its questionnaire about the investor's goals, but how these goals impact the risk-driven investment style is not transparent. A similar picture is given by the other large German robo-advisors: they generally ask in their questionnaires for financial goals, often linked with the time horizon. But an explicitly goal-based investment approach cannot be recognized. Weltinvest even does not ask for goals and does not allow holding multiple portfolios currently.

4.4 Overconfidence

"What do you think: Are you better or worse than the average driver?" If you pose this question to a significant sample of people, most of the time you will probably obtain a result estimating that much more than 50% of the participants appraise themselves as better than the average driver—

which is, in fact, impossible from a statistical point of view.[4] In his book, Taleb (2007) proposes many different questions as to how to test what he calls "epistemic arrogance." With these questions, the issue is not "to gauge [people's] knowledge, but rather their evaluation of their knowledge" (Taleb 2007, p.139). The overwhelming result reported by Taleb (2007) is that people tend to overestimate their knowledge. This overconfidence is a form of illusion of control and a well-documented bias in the scientific literature. It traces back to the works of Tversky and Kahneman (1974), Fischhoff et al. (1977), and Fischhoff et al. (1980), who observed the tendency of people to overestimate themselves in different aspects. This bias is problematic because "overconfidence can keep us from realizing how little we know and how much additional information is needed about the various problems and risks we face" (Slovic et al. 1981). Therefore, overconfidence can be observed in at least three major occurrences (Moore and Schatz 2017):

> Overestimation is thinking that you are better than you are. Overplacement is the exaggerated belief that you are better than others. Overprecision is the excessive faith that you know the truth.

Especially on financial markets, participants are not free from the overconfidence bias: For example, Barber and Odean (2013, p.1547) describe that overconfidence may "explain the relatively high turnover rates and poor performance of individual investors." They refer to papers of Dorn and Huberman (2005), Glaser and Weber (2007), and Grinblatt and Keloharju (2009) to support their hypothesis. Furthermore, as Hirshleifer (2015) points out, on the one hand, overconfident investors tend to reduce diversification in their portfolios; but on the other, they may show over- or underreaction toward prices due to misleading information signals. It is not that private investors alone are prone to overconfidence; professional investors as well seem to fall for this bias. A study by Puetz and Ruenzi (2011) shows that even mutual fund managers tend to increase their turnover rates after experiencing a period of prosperous portfolio performance. After excluding factors such as incentives, inflows, or managerial

[4] This specific example is taken from the German book by Stock and Goldberg (2013): Genial einfach entscheiden, Finanzbuch Verlag. The question, that is, the observation, goes back to Svenson (1981).

learning as potential source for the increased trading activity, they conclude that overconfidence of fund managers is the most likely explanation for their findings. In a conference paper, Choi and Lou (2010) show that in their framework, inexperienced mutual fund managers are more influenced by the overconfidence bias than their more experienced colleagues. In general, studies largely report that overconfidence increases costs and, therefore, decreases portfolio performance, indicating that avoiding overconfidence is beneficial for investors.

4.4.1 How Confident Are Robo-Advisors?

Based on the claim to provide rational investment advice, robo-advisors should base their processes on risk profiling of investors, subsequent and adequate portfolio recommendations, as well as the implementation and monitoring of the suggested portfolio strategies. Active management, which includes the sales of actively managed mutual funds, should be avoided. It is likely that these strategies will only raise costs but will not improve long-term returns.

The large US robo-advisors typically rely on passive investments by applying ETFs and/or index funds. Vanguard, for example, writes that its investment philosophy is based on "benefits of low costs, diversification, and indexing," and it is primarily using stocks, bonds, and cash for its portfolio allocation (Vanguard Advisers, Inc. 2019). Schwab as well prefers ETFs for its asset allocation; however, it also includes more complex asset classes, such as real estate and commodities. Schwab explicitly states that the algorithms its robo-advisor is using are "not designed to actively manage asset allocations based on short-term market fluctuations" (Charles Schwab & Co. Inc. 2019). Wealthfront joins the group by stating that it is "rooted in passive investing, which means we'll build you a globally diversified portfolio of low-cost index funds" (Wealthfront 2020). The asset classes of Wealthfront compare to those of Schwab: stocks, bonds, real estate, and commodities. Betterment does not only rely on ETFs for its allocation of assets such as stocks, bonds, and cash, it also commits itself to "systematic decision-making" (Grealish 2019). Compared to its peers, it is the only robo-advisor out of the big four that applies a rather consequent rule-based investing approach. Grealish (2019) states that Betterment's portfolio optimization is based on modern portfolio theory and the models by Fama and French (1992) and Black and Litterman (1992). In sum, it could be assumed that the big four US robo-advisors avoid by and large the overconfidence bias through the application of passive investment strategies. But, if we take a closer look, Vanguard, Charles Schwab, and Wealthfront do not follow a holistic rule-based approach for their asset allocations but allow expert investment committees to interfere.[5] This is a potential gateway for all kinds of behavioral biases if experts have the possibility of altering the proposed asset allocations. Also, there is the risk to deviate from the trajectory, especially in times of distress—not to mention the biases that could emerge from group decision- making or, more precisely, from hidden profiles or groupthink.

The large European robo-advisors as well as their US peers widely rely on passive instruments such as ETFs. Cominvest and Liqid, however, are exceptions and also offer actively managed mutual funds in their allocations. Moreover, there is a tendency of European robos to be more active in investments than their US competitors. Nutmeg offers two different asset allocation models: fully managed by portfolio managers or fixed allocation. For the fully managed approach, Nutmeg advertises its active investment approach: "This is not a static, one-off process. We continually review the asset allocation for all our customers to decide if we need to make adjustments" (Port 2013). A similar approach is taken by Liqid that offers an active investment style in its "Select" model. Scalable does not call itself "active manager" but applies a mix of passive investment and active risk management based on Value-at-Risk. This seems to be comparable to Vanguard's method, which also promotes active risk management.[6] It is, nonetheless, questionable if portfolio management by active risk adjustments is more successful than a pure buy-and-hold (with rebalancing) approach. Quirion, however, follows a rule-based and passive investment style and, therefore, seems to allow very little or no

[5] Vanguard: "When recommending, setting, and adjusting your asset allocation, we weigh shortfall risk—the possibility that a financial plan or Portfolio will fail to meet longer-term financial goals—against market risk." Schwab: "[W]e have dedicated an entire team of Charles Schwab Investment Advisory (CSIA) experienced analysts to continually use state-of-the-art research and evolve our approach to creating asset allocations designed to improve outcomes for individual investors." Wealthfront: "Wealthfront combines the judgment of its investment team with state of the art optimization tools to identify efficient portfolios."

[6] Vanguard Advisers, Inc. (2019, p.18): "When recommending, setting, and adjusting your asset allocation, we weigh shortfall risk—the possibility that a financial plan or Portfolio will fail to meet longer-term financial goals—against market risk."

human interference. Weltinvest only has fixed asset allocations and does not provide advice on risk profiles of assets, which could lead investors to choose inappropriate portfolios from a risk perspective.

4.5 Summary

Summing up the findings from our analysis, we find that robo-advisory is not completely free from behavioral biases.

Interestingly, the home bias is relatively strong amongst the robo-advisors. Although it is not easy to find "airtight" evidence for the home bias, it is surprising that U.S. robo-advisors clearly have different allocations between foreign and domestic assets compared to their European peers. It is quite evident as all robo-advisors have the tendency to overweight domestic assets. Since many robo-advisors still rely on investment committees and less on rule-based investing, this seems to abet the home bias.

It is noteworthy that the relatively large "non-automatized" portion in the asset management process of robo-advisors opens a gateway to different forms of behavioral biases. For "hybrid" robo-advisors, such as Vanguard's approach, the dependence is more on "emotional sensitivity" than algorithm-based models. With respect to the overconfidence bias, most robo-advisors primarily rely on passive investment instruments such as ETFs or index funds. But even if they largely abstain from actively managed mutual funds, most of them do not seem to believe in simple "buy-and-hold" with regular rebalancing. In a best-case scenario, they apply a (relatively) strict rule-based approach, but in general there are investment committees installed. All of these forms, be it active risk management or active interference in times of turmoil, indicate the underlying belief that active intervention improves the performance of the investment. In sum, the large European robo-advisors do offer more active components in their asset management process than those of their US peers. If we consult the performance analysis in Sect. 1.5 of this book, it can be found that despite the active management elements, robo-advisors do not frequently outperform reasonable benchmarks. So, at least, it seems that robo-advisors as well (or maybe better, their creators) are not insensitive to the overconfidence bias since they charge fees for the additional services provided.

Regarding mental accounting as well, there seems to be a clear difference between US and European robo-advisors. Whereas the US robos align to the goal-based investment approach and hence consider that investors might have different risk tolerances for different investment goals, the European robos seem to ask for goals as it is obligatory. But it does not seem to be deep-rooted in their investment philosophies.

So, is *robo economicus* or *robo sapiens* managing the investor's assets? Although we believe that robo-advisors indeed are capable of helping one make better investment decisions in general, the present situation shows a relatively distinct human intervention in the investment processes. There is, however, a real possibility that completely rational and bias-free robo-investment will emerge in the (near) future.

References

Alan Ahearne, William L. Griever, and Francis Warnock. 2004. "Information Costs and Home Bias: An Analysis of US Holdings of Foreign Equities". *Journal of International Economics* 62(2): 313–336.

Brad M. Barber and Terrance Odean. 2013. "The Behavior of Individual Investors". In *Handbook of the Economics of Finance*, Vol. 2, Part B, Chap. 22, pp. 1533–1570. Elsevier B.V. https://doi.org/10.1016/B978-0-44-459406-8.00022-6.

Fischer Black and Robert Litterman. 1992. "Global Portfolio Optimization". *Financial Analysts Journal* 48(5): 28–43. https://doi.org/10.2469/faj.v48.n5.28.

Jean L. Brunel. 2011. "Goal-Based Wealth Management in Practice". *The Journal of Wealth Management* 14(3): 17–26. https://doi.org/10.3905/jwm.2011.14.3.017.

Charles Schwab & Co. Inc. 2020. *Goal Tracker for Schwab Intelligent Portfolios*. Last Accessed: March 20, 2020. https://intelligent.schwab.com/public/intelligent/insights/whitepapers/goal.html.

Charles Schwab & Co. Inc. 2016. *How Savings "Buckets" Can Help You Reach Your Goals*. Last Accessed: March 20, 2020. https://intelligent.schwab.com/public/intelligent/insights/blog/savings-buckets.html.

Charles Schwab & Co., Inc. 2019. *Schwab Intelligent Portfolios Solutions, Disclosure Brochure*. Last Accessed: September 29, 2019. https://www.schwab.com/public/file/CMS-BDL100049.

Ashvin B. Chhabra. 2005. "Beyond Markowitz: A Comprehensive Wealth Allocation Framework for Individual Investors". *The Journal of Wealth Management* 7(4): 8–34.

Darwin Choi and Dong Lou. 2010. "A Test of the Self-Serving Attribution Bias: Evidence from Mutual Funds". *SSRN Electronic Journal*. https://doi.org/10.2139/ssrn.1100786.

Joshua D. Coval and Tobias J. Moskowitz. 1999. "Home Bias at Home: Local Equity Preference in Domestic Portfolios". *The Journal of Finance* 54(6): 2045–2073. 10.1111/0022-1082.00181.

Sanjiv Das, Harry Markowitz, Jonathan Scheid, and Meir Statman. 2010. "Portfolio Optimization with Mental Accounts". *Journal of Financial and Quantitative Analysis* 45(2): 311–334.

Deutsches Aktieninstitut. 2020. *Return Triangles.* Last Accessed: March 20, 2020. https://www.dai.de/en/what-we-offer/studies-and-statistics/return-triangles.html.

Daniel Dorn and Gur Huberman. 2005. "Talk and Action: What Individual Investors Say and What They Do". *Review of Finance* 9(4): 437–481.

Dan Egan. 2019. *Using Investment Goals at Betterment.* Last Accessed: March 20, 2020. https://www.betterment.com/resources/goals-advice-explained/.

Eugene F. Fama and Kenneth R. French. 1992. "The Cross-Section of Expected Stock Returns". *The Journal of Finance* 47(2): 427–465. https://doi.org/10.2307/2329112.

Michael Fidora, Marcel Fratzscher, and Christian Thimann. 2007. "Home Bias in Global Bond and Equity Markets: The Role of Real Exchange Rate Volatility". *Journal of International Money and Finance* 26(4): 631–655.

Baruch Fischhoff, Asher Koriat, and Sarah Lichtenstein. 1980. "Reasons For Confidence". *Journal of Experimental Psychology: Human Learning and Memory* 6: 107–118. https://doi.org/10.1037/0278-7393.6.2.107.

Baruch Fischhoff, Paul Slovic, and Sarah Lichtenstein. 1977. "Knowing with Certainty: The Appropriateness of Extreme Confidence". *Journal of Experimental Psychology: Human Perception and Performance* 3(4): 552–564. https://doi.org/10.1037/0096-1523.3.4.552.

Hisham Foad. 2011. "Familiarity Bias". *Behavioral Finance: Investors, Corporations, and Markets*, Chap. 15, pp. 277–294. John Wiley & Sons, Ltd. https://doi.org/10.1002/9781118258415.ch15.

Kenneth R. French and James M. Poterba. 1991. "Investor Diversification and International Equity Markets". *American Economic Review* 81(2): 222–226.

Markus Glaser and Martin Weber. 2007. "Overconfidence and Trading Volume". *The Geneva Risk and Insurance Review* 32(1): 1–36. https://doi.org/10.1007/s10713-007-0003-3.

Adam Grealish. 2019. *The Betterment Portfolio Strategy.* Last Accessed: March 20, 2020. https://www.betterment.com/resources/betterment-portfolio-strategy/.

Mark Grinblatt and Matti Keloharju. 2009. "Sensation Seeking, Overconfidence, and Trading Activity". *Journal of Finance* 64(2): 549–578. https://doi.org/10.1111/j.1540-6261.2009.01443.x.

Marcel Grzanna. 2018. *Robos im Aufwind.* Last Accessed: March 20, 2020. https://www.sueddeutsche.de/wirtschaft/robo-advisors-robos-im-aufwind-1.4197562.

Yuval N. Harari. 2015. *Sapiens: A Brief History of Humankind.* Vintage.

David Hirshleifer. 2015. "Behavioral Finance". *Annual Review of Financial Economics* 7(1): 133–159. https://doi.org/10.1146/annurev-financial-092214-043752.

Daniel Kahneman and Amos Tversky. 1979. "Prospect Theory: An Analysis of Decision Under Risk". *Econometrica, Econometric Society* 47(2): 263–291.

Karen K. Lewis. 1999. "Trying to Explain Home Bias in Equities and Consumption". *Journal of Economic Literature* 37(2): 571–608.

Torben Lütje and Lukas Menkhoff. 2007. "What Drives Home Bias? Evidence from Fund Managers' Views". *International Journal of Finance & Economics* 12(1): 21–35. https://doi.org/10.1002/ijfe.309.

Don A. Moore and Derek Schatz. 2017. "The Three Faces of Overconfidence". *Social and Personality Psychology Compass* 11(8). https://doi.org/10.1111/spc3.12331.

Nutmeg. 2020. *How We Invest.* Last Accessed: March 20, 2020. https://www.nutmeg.com/how-we-invest#portfolio-performance.

Carrie H. Pan and Meir Statman. 2012. "Questionnaires of Risk Tolerance, Regret, Overconfidence, and Other Investor Propensities". *SCU Leavey School of Business Research Paper*, 10-05.

Shaun Port. 2013. *An Inside View of the Nutmeg Investment Process: Asset Allocation.* Last Accessed: March 20, 2020. https://www.nutmeg.com/nutmegonomics/an-inside-view-of-the-nutmeg-investment-process/.

Alexander Puetz and Stefan Ruenzi. 2011. "Overconfidence Among Professional Investors: Evidence from Mutual Fund Managers". *Journal of Business Finance & Accounting* 38(5-6): 684–712. https://doi.org/10.1111/j.1468-5957.2010.02237.x.

Chad Rixse. 2018. *The 4 Advantages of Human vs. Robo-Advisors.* Last Accessed: March 20, 2020. https://www.cnbc.com/2018/4/25/the-4-advantages-of-human-vs-robo-advisors.html.

Scalable Capital. 2018. *Scalable Capital gewinnt Openbank die Digitalbank der Santander- Gruppe, als Partner in Spanien.* Last Accessed: March 20, 2020. https://de.scalable.capital/presse/scalable-capital-gewinnt-openbank.

Mark S. Seasholes and Ning Zhu. 2010. "Individual Investors and Local Bias". *Journal of Finance* 65(5): 1987–2010.

Hersh Shefrin. 2000. *Beyond Greed and Fear: Understanding Behavioral Finance and the Psychology of Investing.* Harvard Business School Press.

Paul Slovic, Baruch Fischhoff, and Sarah Lichtenstein. 1981. "Facts and Fears: Societal Perception of Risk". *Advances in Consumer Research* 8: 497–502. http://www.acrwebsite.org/search/view-conference-proceedings.aspx?Id=5844.

Christin Stock and Joachim Goldberg. 2013. *Genial Einfach Entscheiden: Besser Denken, Handeln und Investieren im Täglichen Entscheidungsdschungel.* FinanzBuch Verlag.

Ola Svenson. 1981. "Are We All Less Risky and More Skillful than our Fellow Drivers?" *Acta Psychologica* 47(2): 143–148. https://doi.org/10.1016/0001-6918(81)90005-6.

Nassim Taleb. 2007. *The Black Swan: The Impact of the Highly Improbable.* Random House Group.

Michael Tertilt and Peter Scholz. 2018. "To Advise, or Not to Advise— How Robo- Advisors Evaluate the Risk Preferences of Private Investors". *The Journal of Wealth Management* 21(2): 70–84.

Linda L. Tesar and Ingrid M. Werner. 1995. "Home Bias and High Turnover". *Journal of International Money and Finance* 14(4): 467–492. https://doi.org/10.1016/02615606(95)00023-8.

Richard Thaler. 1985. "Mental Accounting and Consumer Choice". *Marketing Science* 4(3): 199–214.

Richard Thaler. 1980. "Toward a Positive Theory of Consumer Choice". *Journal of Economic Behavior & Organization* 1(1): 39–. https://doi.org/10.1016/0167-2681(80)90051-7.

Harish Thiagarajan. 2018. *Testing the GPS Behind Tomorrow's Self-Driving Money.* Last Accessed: March 20, 2020. https://eng.wealthfront.com/2018/06/28/testing-the-gps-behind-tomorrows-self-driving-money/.

Amos Tversky and Daniel Kahneman. 1974. "Judgment under Uncertainty: Heuristics and Biases". *Science* 185(4157): 1124–1131. https://doi.org/10.1126/science.185.4157.1124.

Amos Tversky and Daniel Kahneman. 1981. "The Framing of Decisions and the Psychology of Choice". *Science* 211: 453–458. https://doi.org/10.1126/science.7455683.

Vanguard. 2020. *Introducing Vanguard Personal Advisor Services.* Last Accessed: March 20, 2020. https://investor.vanguard.com/financial-advisor/personal-advisor-services.

Vanguard. 2020. *Investing Goals.* Last Accessed: March 20, 2020. https://investor.vanguard.com/investing/goals.

Vanguard. 2020. *Vanguard Personal Advisor Services.* Filmed October 14 2015 at Youtube. Video 1:52. https://www.youtube.com/watch?v=OYgJhCoAUnA.

Vanguard Advisers, Inc. 2019. *Vanguard Personal Advisor Services Brochure.* Last Accessed: September 29, 2019. https://personal.vanguard.com/pdf/vpabroc.pdf.

Wealthfront. 2020. *Investing Isn't an Art. It's a Science.* Last Accessed: March 20, 2020. https://www.wealthfront.com/investing.

CHAPTER 5

Quant Models for Robo-Advisors

Thorsten Ruehl

Contents

T. Ruehl (✉)
Head of Investment Strategies and Research, CSR Beratungsgesellschaft mbH, Hofheim am Taunus, Germany
e-mail: thorsten.ruehl@csr-beratungsgesellschaft.de

© The Author(s) 2021

P. Scholz (ed.), *Robo-Advisory*, Palgrave Studies in Financial Services Technology, https://doi.org/10.1007/978-3-030-40818-3_5

5.1 Introduction

One of the main advantages of robo-advisory is the ability to offer a large number of investors automated and thus cost-efficient asset management that can still be tailored to the client's needs to a certain extent. The easy scalability combined with high individuality (compared to the still-dominant standard solutions for retail customers) is one of the great strengths of robo-advisory solutions (Bankenverband 2017). To take full advantage of these benefits, all the components of a robo-advisory platform must work together as effectively as possible. This, in turn, places demands on investment strategies that can be used in this context. It makes sense to favor concepts that are both automated and rule based and which can be easily parameterized to reflect individual client requirements. This is the only way to derive full benefit from the scaling advantages offered by robotics (Lam 2016).

When comparing robo-advisors with independently managed security accounts or (conventional) professional investment advisors, the benefits typically mentioned include the following: *low costs*, *focus on risks*, *technology instead of emotions*, and *transparency*. All four points can be attributed to automation benefits: As in other industries, automation also leads to a reduction in costs, as labor now represents the largest cost item in many areas. Risks, in turn, can only be quantified and controlled by financial mathematical models and calculations. This process is inherently linked to the use of computers and thus predestined to be part of robo-advisory.

The consistent and systematic adherence to an investment approach is significantly facilitated by a purely rule-based and thus technically mappable approach. The typical errors in investor behavior can also be greatly mitigated by the systematic use of smart, stringent approaches. One of the frequently observed but avoidable investor mistakes is to exit from a long-term successful systematic approach "at an inopportune time". An investment strategy that is understood and "supported" in its decisions thus helps investors to stay "on board" even in difficult market phases and to avoid logging in losses. Experience has shown that this advantage of a strictly rule-based and transparent investment strategy is often underestimated. If the investor has understood the basic rules of the investment strategy, he will be able to understand the strategy's behavior (and is outcome) in different market phases and will therefore be able to stick with the strategy even in difficult times, which in turn is important for the success of the investment in the long term. All four of these advantages originate from a common source and can ultimately be traced back to a largely automated use of algorithm-based investment decisions. Since robo-advisory is based on highly automated software platforms, it would be downright wasteful not to take advantage of the resulting benefits at the heart of the investment strategy itself (Lam 2016). We will, therefore, next take a closer look at this aspect.

5.2 What Strategies Are Suitable for Robo-Advisory?

To capitalize on the benefits offered by the technology, investment strategies are required that integrate seamlessly with existing technology and have the same structural benefits. From this point of view, purely quantitative strategies form part of a robo-advisory as the entire process can then be designed "from a single source". In principle, discretionary investment strategies can also be successfully used in asset management. Discretionary strategies, on the other hand, move between the following two poles, that is, hybrids are also possible:

Individual
Every portfolio manager makes investment decisions solely for "his" portfolios. This can mean that portfolio manager A increases the equity allocation at a given point in time, for example while his colleague B reduces it on the same day. From the company's perspective, this approach offers a considerable advantage: the diversification resulting from this organizational structure reduces the likelihood that all portfolios will perform poorly at the same time and lead to overall client dissatisfaction with the risk of concentrated cash outflows. However, a certain herd mentality of the formally independent fund managers cannot be ruled out even with this form of organization, as it is well known that it feels more comfortable to wander with the masses than to wander alone. Only purely quantitative processes are immune to such emotional appraisal processes.

In-House Strategy
An investment committee sets guidelines for the currently supported investment allocation, right up to uniform model portfolios, which must be implemented by all in-house fund managers. Diverging performances

of individual portfolios from the same company are thus avoided, but so are the benefits of style diversification.

The short outline of the two approaches already shows that in the context of a robo-advisory, only the second variant would be considered, if at all: uniform model portfolios that serve to control individual securities accounts by mapping them one-to-one. Although such an approach can actually be implemented, it is considerably more expensive than a purely quantitative solution, since the investment process used for controlling would have to be set up in an entirely discretionary manner, with all the associated disadvantages on the cost side. The cost advantages resulting from a larger number of target portfolios per sample portfolio are already being used today in asset management for smaller portfolios. The last step toward automation is no longer being taken here. However, fund-linked asset management with a small number of discretionarily managed funds of funds, to which the client portfolios are then allocated based on risk appetite, are already consistent with the solution outlined above. When "porting" to a robo-advisor platform, only the front end to the client would change: the investment advisor who makes the selection on behalf of the client on the basis of a predefined list of criteria would be replaced by the robo-platform. A further disadvantage of the discretionary solution approach lies in the limited transparency: although the investment decisions of the investment committee can be published, a uniform approach across all times and personnel changes cannot realistically be guaranteed. This circumstance will sooner or later have a negative impact on portfolios with very long-term horizons, for example for retirement provision purposes, since the investor's reasons for deciding on a certain model portfolio may have become obsolete over the years due to changes in the discretionary process. The disadvantages of discretionary approaches in the context of robo-advisory are that the degrees of transparency, continuity, and cost-efficiency that can be achieved with quantitative approaches can never be fully achieved. Quantitative approaches with transparent rules show their strengths precisely here (Satchell 2003): once an algorithm has been set up, it only requires comparatively inexpensive maintenance at runtime, while the discretionary approach relies on the ongoing work of a (cost-intensive) investment committee. As quantitative rules consist of a fixed, always identical set of rules, they can be made transparent to the investor to an arbitrary degree. Only copy protection, which is not achievable by law, will set limits here in practice, but not the investment strategy per se. An investment strategy that, for example, is decidedly aimed at seeking the same risk contributions per asset class at all times will continue to do so even after 10 or 20 years, that is, the advantage in terms of continuity, in addition to the transparency advantage, results in a type of "accompanying advantage" over the discretionary approach. We will therefore deal with purely quantitative approaches and identify those particularly suitable for use as part of robo-advisory.

5.3 What Quantitative Approaches Does the Robo-Advisory Model Offer?

Robo-advisory services take advantage of automated processes—it is important to pursue this idea consistently right down to the investment strategy. However, not every rule-based approach is equally suitable for use in a fully automated implementation. There are many technical approaches that evaluate historical price patterns and draw conclusions about the current market situation. A simple example would be the use of moving averages to determine entry and exit times for any given market (Brock et al. 1992). Such approaches can be very successful in practice. However, they are not based on a strictly scientific basis, but on the use of a (mathematically formulated) heuristic. In order to do this, the so-called back tests are carried out, but their prognostic significance or temporal stability is often not given. Here, too, "post-optimization" must be carried out on an ongoing basis in the future, at the strategic level rather than at the portfolio level. These strategies thus come close to discretionary strategies, with all the advantages and disadvantages already mentioned above, especially in terms of transparency and continuity. In order to take full advantage of the aforementioned options that are available in robo-advisory, the strategies used must therefore also have a high degree of stability over time (Meucci 2009; Grinold and Kahn 2012). Risk models satisfy this requirement, while return forecasting models have to be revised regularly and they are also disadvantageous in terms of the required transparency.

For this reason, we want to focus primarily on quantitative approaches that make do with pure risk management and do not include return forecasts in the optimization process. What strategies fall into this category? First of all, these are all quasi-stationary strategies in which only a regular rebalancing (practical adjustment frequencies range from weekly to annual) is carried out according to a fixed rule. This fixed rule can be, for example, an equal weighting or a weighting based on market capitalization.

Such strategies have been devised to be highly transparent and easy to implement. The latter point, however, ensures that these strategies can also be easily "replicated", and thus they will always be subject to increased price pressure. From the provider's perspective, more sophisticated approaches should not only be aimed at benefiting the investor. Thanks to the automated platforms, however, such more sophisticated approaches can be implemented with comparatively little additional cost. Added value for the client can be achieved, for example, through the following objectives:

- Maximization of the diversification effect
- Equal distribution of risks to the investment instruments contained in the portfolio
- Risk minimization
- Adherence to lower value limits
- Specification of a risk preference by choosing a target investment period

5.3.1 Maximization of the Diversification Effect

The old stock market wisdom of not putting all your eggs in one basket is often cited, but too often not consistently followed. To be clear: Many baskets are also of little use if they are mounted on the same bike rack and the whole bike tips over. It will be difficult to achieve a noticeable stabilization of the portfolio through diversification effects with equities from a single sector. A necessary but not yet sufficient prerequisite for a well-diversified portfolio is, therefore, an investment universe that not only consists of highly correlated components, but also makes targeted use of those with low or even negative correlations. The more the components differ, the greater is the chance that even in times of crisis the portfolio can be effectively hedged through opposing developments. Such a well-diversified investment universe is, therefore, also a necessary prerequisite for the construction of risk-controlled portfolios on robo-advisor platforms. However, this is only half the battle. The main audience for robo-advisor platforms are private investors who are unlikely to have any experience with investment mathematics. This is where the robo-advisor platform can demonstrate its strengths, for example, by determining the mixing ratio of the components for the available part of the investment universe, where the diversification effect is greatest. This approach can be formulated mathematically and transformed into an optimization problem (Choueifaty and Coignard 2008). All you have to do is maximize the diversification ratio *DR*, which can be determined as follows:

$$\max_{w_i} DR = \max \frac{\sum_{i=1}^{N} w_i \cdot \sigma_i}{\sigma_p} \tag{5.1}$$

given the boundary conditions

$$\sum_{i=1}^{N} w_i = 1$$

and

$$w_i \geq 0 \quad \forall \quad i = 1, \ldots, N$$

with w_i as weight of asset *i* within the portfolio, σ_i as volatility of asset *i*, σ_p as portfolio volatility, and *N* as number of assets. Or, put in another way, the diversification ratio can be expressed as portfolio risk without diversification divided by portfolio risk with diversification. Hence, the weighted sum of asset risk divided by the total portfolio risk equals the maximum diversification ratio *DR* at its peak. In this type of portfolio optimization, the ratio of the weighted individual risks of the asset classes (excluding diversification) to the actual portfolio risk (i.e. including diversification) is maximized. This process in the two-asset case can be illustrated as displayed in Fig. 5.1. If the two axes are swapped and the diversification ratio is also plotted, then the point you are looking for in the graph can be read directly as the maximum (see Fig. 5.2). The advantage of a portfolio structured in this way is that it has the highest risk-adjusted diversification effect of all portfolios that can be built from the investment universe (Choueifaty et al. 2013). Elements from the investment universe that diversify well are highly rewarded, even if, in themselves, they might not have been considered when using other portfolio structuring techniques (such as variance minimization) due to their volatility, which may be somewhat higher.

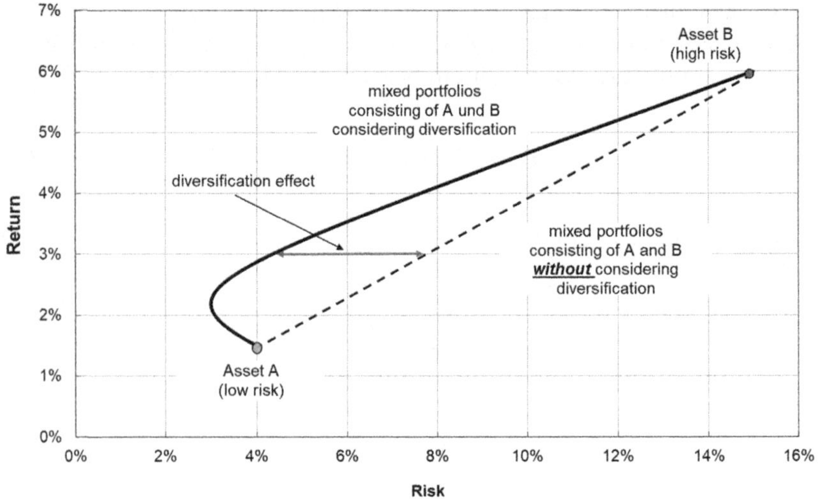

Fig. 5.1 Diversification effect

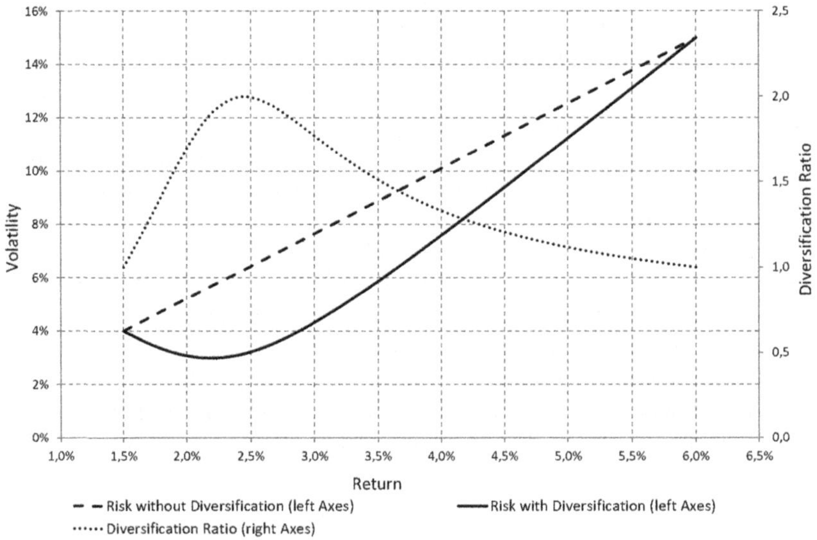

Fig. 5.2 Diversification ratio shows maximum at 2.0

5.3.2 *Equal Distribution of Risks to the Investment Instruments Contained in the Portfolio*

The maximum diversification approach described above not only optimizes the allocation weightings based on this target, but also implicitly selects the investment instruments from the investment universe, that is, not all available instruments are necessarily included in the portfolio. This can frustrate some investors who have actively chosen a number of instruments and are now disappointed not to find them all in their portfolio. In this case, there is a way to ensure that all previously selected instruments are actually included in the portfolio, while maintaining a balance between them in terms of risk. This method therefore assigns the same risk contribution to all instruments that are to be found in the portfolio. Portfolios built in this way have become quite popular in recent years and are referred to as "risk parity" portfolios. With risk parity, the portfolio is optimized in such a way that all instruments have the same contribution to the total risk (Teiletche et al. 2010).

$$PCTR_i = \frac{ACTR_i}{\sigma_p} = \frac{1}{N} \quad \forall \quad i = 1, \ldots, N \tag{5.2}$$

where N is number of assets, σ_p is portfolio volatility, $ACTR_i$ is absolute contribution to total risk of asset i, and $PCTR_i$ is percentage contribution to total risk of asset i. The advantage of a portfolio built in this way is that it avoids structural cluster risks. The correlations between the asset classes and thus their diversification potential are explicitly taken into account. However, during major financial market crises, the effect can be observed time and again that investors on a large scale close out risky positions across markets and regions and withdraw liquidity from the market. As a result, the correlations between these risky instruments rise abruptly during the crisis (so-called diversification breakdown). In other words, where there was protection by diversification at least on paper, when it is needed most urgently, it is gone.

In order to anticipate this effect of the increasing correlations in the crisis, a modified approach can therefore be chosen in advance, in which uniform volatility contributions are allocated instead of uniform risk contributions. This modification not only protects against unpleasant surprises during market corrections, but also offers the advantage of easier computation, as the portfolio weights can be calculated directly without having to carry out a (more time-consuming) optimization.

$$w_i \cdot \sigma_i \equiv w_j \cdot \sigma_j \quad \forall \quad i, j \Leftrightarrow w_i = \frac{\frac{1}{\sigma_i}}{\sum_{j=1}^{N} \frac{1}{\sigma_j}} \quad \forall \quad i = 1, , ..., N \tag{5.3}$$

with N as number of assets, w_i as weight of asset i, and σ_i as volatility of asset i. The latter modification retains the advantage of taking into account all preselected instruments from the investment universe. This also simplifies the calculation compared to the conventional risk parity method (as no optimization is required) and makes the portfolio less sensitive to a "diversification breakdown" at times of crisis.

5.3.3 Risk Minimization

The two purely quantitative approaches described above ensure that either the potential diversification effect is fully exploited for a given investment universe or that the risks contained in the portfolio are distributed as evenly as possible. For particularly risk-averse investors, however, it is advisable to make portfolios available that minimize the overall portfolio risk (Clarke et al. 2011):

$$\begin{aligned} \min_{w_i} \sigma_p^2 &= \min \sum_{i=1}^{N} \sum_{j=1}^{N} w_i \cdot w_j \cdot \sigma_{i,j} \\ &= \min_{w_i} \underbrace{\sum_{i=1}^{N} w_i^2 \cdot \sigma_i^2}_{\text{single risk part}} + \underbrace{\sum_{i=1}^{N} \sum_{\substack{j=1 \\ i \neq j}}^{N} w_i \cdot w_j \cdot \sigma_{i,j}}_{\text{diversification part}} \end{aligned} \tag{5.4}$$

with boundary conditions

$$\sum_{i=1}^{N} w_i = 1$$

and

$$w_i \geq 0 \quad \forall \quad i = 1, ..., N$$

as well as N as number of assets, σ_p^2 as portfolio variance, $\sigma_{i,j}$ as covariance of asset i and j, and w_i as weight of asset i.

Portfolios built in this way (generally based on factor models) have become popular in recent years due to their outperformance in equities. In theory, however, this portfolio has a serious disadvantage: it is "below" the capital market line, that is, the combination of a tangency portfolio and the risk-free investment leads (theoretically!) to the same low risk with higher expected returns. In practice, however, this disadvantage can be ignored: as return forecasts are needed to build a tangency portfolio. However, it is not possible to use historical data to come up with even short-term forecasts with the same confidence as it is the case with pure risk indicators. A tangency portfolio calculated using mean-variance optimization is, therefore, fraught with such uncertainty that the minimum variance portfolio, figuratively speaking, still lies within the error bars. The disadvantages that are relevant in practice are of a different kind: as with the portfolio with the maximum diversification ratio, even the minimum variance portfolio does not ensure that all previously selected instruments are included in the portfolio. In addition, the minimum variance optimization in mixed portfolios generally leads to a very high share of bonds, because it takes into account their low volatility, but not their low continuous yield.

5.3.4 *Methods Based on Return Forecasts*

The methods presented so far are purely risk based, that is, the expected return forecasts were deliberately omitted in order to circumvent the associated forecasting error issues. Even if forecast-free strategies have very advantageous characteristics despite the complete renunciation of an assessment of the market movement and have already been able to hold their own on the market (Clarke et al. 2013), it could, nevertheless, be argued that this in a way throws the baby out with the bathwater, because to avoid the problems associated with return forecasts, these have been completely foregone. In the following, therefore, we show a viable path to conventional portfolio optimization, in which a mean variance optimization (MVO), according to Markowitz, is performed using expected returns (Markowitz 1952):

$$U = \mu_p - \lambda \cdot \sigma_p^2 \tag{5.5}$$

where μ_p is expected portfolio return, σ_p^2 is portfolio variance, and λ is the risk aversion parameter.

The uncertainty under which the MVO is optimized takes into account only the dispersion of market returns, reflecting portfolio volatility σ. The return estimates for the individual portfolio components, which aggregate the expected portfolio return μ_p, are, however, implicitly assumed to be the exact mean of the distribution. This assumption, which is far removed from practice, leads to some very undesirable effects in portfolios optimized in this way: for example, in an MVO, highly correlated assets are considered perfect substitutes and are played "against each other" due to small differences in the return estimate, although the actual forecast error may be of the same magnitude as the estimated return spread. In other words, what at first glance looks like taking advantage of an arbitrage opportunity may turn out to be merely reinforcing a forecasting error ex post. Over time, comparatively small changes in the return estimates, which in reality, are due to forecasting errors, can lead to allocation leaps that ultimately rely on artifacts. These disadvantages of the MVO can be mitigated by a suitable transformation of return estimators. The Black-Litterman model (Litterman 2003) is, for example, very suitable for this purpose. The Black-Litterman model supplements the pure MVO with a process step in which the "raw" return estimators are modified as follows:

- Forecasts for highly correlated markets will be aligned based on this information. This counteracts the MVO's ability to treat highly correlated assets as perfect substitutes from a risk perspective and to "play them off" against each other in the event of diverging forecasts.
- Forecasts with higher confidence are given a greater consideration than those with lower confidence. This approach is intuitive. Borderline cases are pure MVO (all forecasts are highly reliable) and a preselected anchor portfolio (e.g. the investor's long-term benchmark portfolio) in the event that no reliable forecasts are available. In turn, the forecast-free approaches outlined above can be used as an anchor portfolio, so that in the event of high forecasting uncertainty an allocation that is advantageous from a pure risk perspective can be targeted.

The forecasting quality (confidence) can be determined by a sliding measurement of the variance of the forecasting errors, implicitly assuming a certain persistence in the quality of the estimates.

A fundamental disadvantage of forecast-based models, as mentioned at the beginning, is that return forecast models are associated with increased maintenance costs compared with pure risk models. However, this can be limited by picking the right model. Based on our own experience, the general regression neural network (GRNN), which is an extension of the probabilistic neural network (PNN) for non-discrete allocations, is very well suited, as it allows GRNNs to be used to approximate non-linear correlations such as price forecasts based on economically relevant variables. The GRNN is based on a very intuitive basic assumption: the more similar the past explanatory variables are to the current constellation, the more likely it is that the following price performance will closely resemble past performance. Another advantage of the GRNN is that only a single free parameter needs to be determined by optimization. This is the size of the neighborhood in the weighted approximation. If the chosen neighborhood parameter is infinitely large, on the one hand, an arithmetic averaging overall historical events will result. If the chosen neighborhood parameter, on the other hand, is infinitely small, the GRNN will simply act as a nearest neighbor estimator. Realistic neighborhood settings will of course lie between those two extremes.

The GRNN is ideally suited for adaptive forecasts, as each new input vector (consisting of the currently measured relevant economic variables) with the corresponding realized market return can be easily integrated into the existing database, and thus can be immediately fed into the next return estimate. This approach, therefore, has considerable advantages for robo-advisory services discussed here. With conventional regression analysis—if one wanted to use such an adaptive method—it would be necessary to reestimate the regression coefficients on an ongoing basis or to redevelop the regression function for every newly added data set completely from scratch, which would be even more time consuming. A detailed description of the procedure can be found, for example, in Specht (1991) and Rühl (2001).

5.4 Dealing with Risk Targets

5.4.1 *Adherence to Lower Value Limits*

The focus on and assessment of risks is rightly considered to be one of the benefits of robo-advisory services. The strategies presented above focus on the risk side: either by avoiding cluster risks, by maximizing the diversification effect, or by minimizing the overall risk. The possibility of automating a robo-advisor platform makes it possible to agree an individual lower value limit at the securities account level. Although lower value limits in the sense of capital preservation are no longer possible for a one-year time period due to the current interest rate environment, a previously accepted loss in value of max. 10%, for example, still represents a considerable limitation of the loss potential compared with an unsecured investment.

The maximum loss on the paid-up capital borne by the investor must be converted into an actual current maximum loss, which takes into account the previous market performance, that is, a positive market performance will increase the actual buffer available, while a negative market performance will erode part of the buffer. The buffer actually available on the basis of these two effects (initial buffer + market performance) then defines the maximum still acceptable value at risk (VaR_{max}) of the portfolio. If the actual value at risk (VaR_{act}) threatens to exceed the remaining buffer, the portfolio will have to become more defensive. The decision-making and control process that must be carried out continuously (and automatically!) in such a portfolio is as follows:

- As long as $VaR_{akt} \leq VaR_{max}$ applies, the current allocation can be retained or any safety measures can be resolved until $VaR_{akt} = VaR_{max}$ again.
- If, however, as a result of a negative performance or an increase in market risk with $VaR_{akt} > VaR_{max}$, a more defensive allocation must be selected until $VaR_{akt} \leq VaR_{max}$ again. The VaR reduction can be achieved either by adding liquidity or by choosing a more defensive but still fully invested allocation.

However, this procedure has one distinct disadvantage: the lower the available risk buffer, the higher is the probability that the portfolio will have to be completely removed from all risky investments (the so-called cash lock). Especially with long investment periods, it can happen that while the investment period, which is still available for value growth, can no longer be used, the portfolio remains "logged in" to the maximum loss. In other words, although it is technically feasible and even practicable at a reasonable cost to protect the value of an individual securities account, it raises the question of what to do in the case of a cash lock. In the case of conventional securities accounts, where clients have access to advisors, a solution can be found through dialogue. However, automated solutions must be offered as part of robo-advisory services. In the case of a hedging horizon of one year, a cash lock can, of course, be "paused" until a new risk buffer is made available again at the beginning of a new calendar year. If, however, a market reset occurs very early in the year that forces the portfolio completely out of the market and the market subsequently recovers, this usually leads to a high disappointment potential, as a negative portfolio result is offset by a positive annual financial statement at a market level. Other forms of risk management are available to avoid such potential conflicts going forward. For a practical implementation of risk targets, for example, an investment period to be chosen by the investor can be specified, at the end of which the invested capital is preserved with a sufficiently high degree of confidence. This takes advantage of the fact that the risk increase will be sharper than linear in the shorter term, but weaker in the long term (Danielsson and Zigrand 2006).

5.4.2 Specification of a Risk Preference by Choosing a Target Investment Period

Instead of working with a maximum loss target, investors can alternatively choose an investment horizon after which the invested capital is highly likely to be maintained at least nominally with a specified level of confidence (e.g. 95% or 99%). Over this period, the expected return of the portfolio "applies" and the expected value after this period is well above 0%. This type of risk target means that the investor is not confronted with very technical specifications such as the choice of a risk aversion parameter λ or a target volatility σ and can focus on the essential: a savings target in the future. Compared to a lower value limit for shorter periods, the advantage of this approach is that it is immune to the cash lock risk.

The mathematical-technical implementation of this specification takes place where it can be solved with comparatively little (additional) cost due to the existing infrastructure: on the robo-advisor platform. Even if short-term return forecasts are extremely unreliable, estimates of excess returns at the asset class level can be made, at least in the long term, with sufficient confidence for the purpose intended here. In order to calculate the required investment horizon for a certain allocation, one takes advantage of the effect that the risk of an investment will increase with the square root of time, that is, $\sigma \sim \sqrt{t}$, and it is thus stronger than linear in the short term, but weaker in the long term.

If one makes a conservative assumption (neglecting the compound interest effect) of a linear increase in the expected return over time, you can determine the point of intersection for each selected allocation and thus the investment period from which you can expect a capital preservation. This can be done for any confidence level using the z-factor of choice. If the investment strategies that can be mapped using the robo-advisory platform are categorized based on their risk/return profiles, it is possible to filter out from this strategy universe those that achieve the sufficiently high confidence point before or at the end of the desired investment horizon. Alternatively or additionally, in the case of mean-variance-optimized strategies, those risk aversion parameters which satisfy this condition can be determined.

5.5 Return Targets and Risk-Bearing Capacity: Need for Information

In discussions with investors, it repeatedly becomes clear that the full implications of the low interest rate environment for the return expectations of all asset classes are all too often insufficiently understood. As a result, there are often unrealistically high expected returns on the one hand and an inappropriate risk bearing capacity on the other. This dual distortion of expectations results in a high potential for disappointment. Anyone who pursues a 5% return target and believes they can do so risk-free in today's capital market environment will almost inevitably be disappointed. It is therefore necessary to provide information about the fundamental relationship between risk and return, which extends beyond regulatory requirements. This also includes fundamental cause–effect relationships. This basic understanding then helps in the selection of the investment strategies or their risk characteristics that are suitable for one's own investment needs.

Because risk premiums are paid relative to risk-free interest, this means that if risk premiums remain the same, total return expectations must decrease as the risk-free rate falls across all premium sources. This can only be compensated by an increased risk premium. According to our own experience, this is, however, not the case in the current environment. The low interest rate environment, therefore, not only affects money market-related forms of investment, but also lowers the realistic expected returns across all asset classes. Unfortunately, this is only one side of the coin. Due to the lower expected value, the return distribution as a whole shifts "to the left", that is, further into the negative range so that all percentiles in the negative range are more likely to occur. In other words: while the expected returns must be adjusted downward, the risk ratios have to be adjusted upward. These two factors should not be lost on new robo-advisor clients, who have made their last investment decision "some time ago".

Actively managed investment strategies may, under favorable circumstances, generate up to one percentage point of additional return for each percentage point of volatility, as a premium to withstand fluctuations. In the following, we will assess a strategy with an expected return of 2.5% p.a. and volatility of 3.0%. Even under this optimistic premise, the risk-bearing capacity required to maintain a confidence level of 99% for a return target of 2.5% p.a. is just under −4.5% on a one-year view. If a 95% confidence is sufficient, the risk-bearing capacity drops to −2.4%, but it is statistically exceeded every 20 years. For example, in the case of a ten-year government bond, the period during which the risk falls to zero means that after ten years, the bond is fully repaid. In the case of actively managed asset management with open maturities, no such risk-free time in the future can be identified, but as outlined above, one can statistically calculate the time after which the paid-in capital is retained or available again with a given confidence. This is done by taking advantage of the risk growing weaker over time (with normally distributed returns proportional to a root function of time) and then determining the time when the expected return will most likely exceed the risk. In order to retain the invested capital with a high probability and a return target of 2.5%, the investment horizon must be extended well beyond one year. In the above example, if the probability of loss over the course of a year is still around 20%, it will fall to just under 12% after two years and to around 3% after five years. After 7.8 years, the loss probability will only be 1%. However, government bonds with this residual maturity have a "guaranteed" negative yield. Figure 5.3 shows the relationship between confidence levels (90%, 95%, and 99%) and minimum

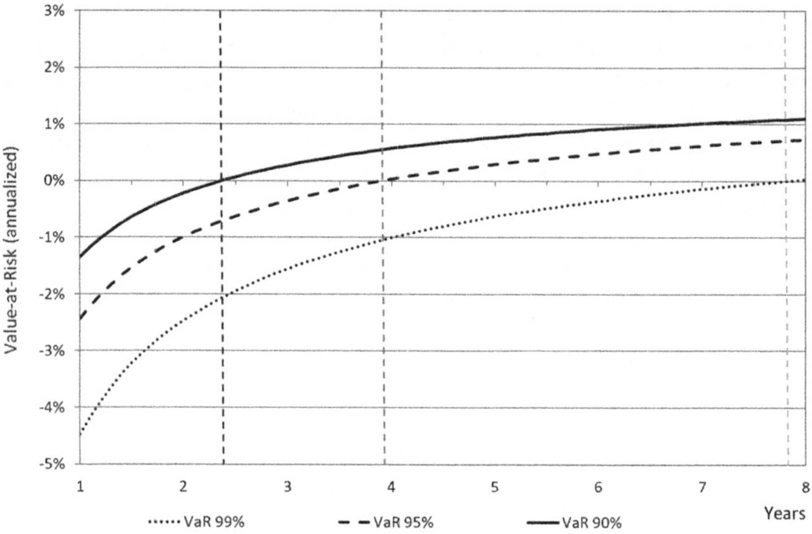

Fig. 5.3 Risk profile of a portfolio with 3% volatility and 2.5% expected return

investment duration for a defensive investment strategy (assumption: 3.0% volatility and 2.5% expected return): The solid line intersects with zero already after 2.4 years, that is, after this time the value at risk has dropped to 0 at 90% confidence level. It will take the aforementioned 7.8 years for the dotted line (99% confidence level) to intersect with 0.

To present these relationships to the (potential) investor at an early stage will also be worthwhile in the long term from the provider's point of view. While this may "put off" a few prospective clients in the short term, in the long term it will ensure a stable client relationship, as this was not entered into under the premise of unrealistically optimistic assumptions.

5.6 Requirements for the Investment Universe and Instruments

The requirements for the investment universe inevitably arise from the points already discussed. The investment universe must allow for sufficient diversification so that strategies focused on risk management can leverage their strengths. In addition, preference should be given to markets that can be modeled with sufficient precision using the risk models used on the platform. For volatility-based risk models, this means that market returns need to be approximated as normally distributed, which can be assumed in many liquid markets if the data frequency is not too high. Our own calculations have shown, for example, that in the current environment corporate bonds can be described with sufficient accuracy using a parametric value-at-risk, up to a confidence level of about 97%. However, in the case of higher confidence levels, the risk is increasingly underestimated when assuming normally distributed returns. In the case of a volatility-adequate mixture of equities and government bonds, the risk can be adequately estimated at the same data frequency with a confidence level of 99% with a (normal distribution-based) parametric value-at-risk.

In principle, it is possible to use simple (plain vanilla) components as part of a robo-advisory, which are then "refined" by using the relevant investment strategy. Although this means abandoning alpha at the component level. However, this potential disadvantage is offset by the fact that passive components are not exposed to the dangers of a manager change and can thus be modeled with a greater degree of confidence. In order to avoid inducing any further avoidable transactions such as rolling transactions other than those induced by the investment strategy, exchange-traded funds (ETFs) are preferable to derivatives despite slight cost disadvantages.

5.7 Customization by Investors

Robo-advisory requires a certain prior understanding on the part of the investor, but can use similar questions as conventional asset managers to lead the investor to the best possible solution. The investor should be able to select the following features of the portfolio or investment strategy:

- *Investment universe*: From the portfolio of available markets, the investor must be able to choose the markets or ETFs to be included (or excluded). To facilitate the selection process for less experienced investors, it is appropriate to define standardized solutions, for example German, Eurozone, European or world equities (or bonds).

If different strategy concepts are offered, this also applies analogously at the strategy level. As a general rule: If forecast-based strategies are used, they will generally operate on a narrower investment universe than the so-called forecast-free strategies. For this reason, the entire robo-platform investment universe will not be available for every investment strategy. To avoid overwhelming inexperienced investors, reasonable standard solutions should also be offered here that reflect the risk categories “conservative” to “aggressive”. This can either take the form of several portfolios graded by risk (conservative, balanced, aggressive) or two basic portfolios at the extreme ends of the risk spectrum (conservative and aggressive), which are then combined to match the investor’s chosen risk profile. This takes us to the next point.

- *The risk appetite* (see also “Risk targets over investment horizons”): Very few investors are able to specify their risk aversion parameters λ for a Markowitz-based optimization. A large proportion of the target audience of a robo-advisory platform will find it difficult to name a concrete target volatility.

Matters are complicated by the following effect, which can be observed quite often: Depending on the current market environment, strategies are often preferred that significantly overburden the investor’s actual risk-bearing capacity in times of crisis. If such a risky strategy is in the immediate vicinity of a new all-time high, its inherent risk is typically underestimated, as “everything has always worked out fine”. If the historical drawdowns, which can be recognized from the graphically visualized time series, are then experienced in real time, they are perceived as much more threatening: What if things do not work out fine this time? Often the investment is then terminated in an untimely manner, and the resulting loss is realized.

It is, therefore, more effective if the investor either specifies the maximum loss amount or specifies the investment horizon according to which at least the capital employed is highly likely to be obtained (or recovered). With the help of these specifications, those strategies can then be presented for further selection along with their risk characteristics that meet these conditions with a high degree of confidence. Depending on the complexity of the platform and the level of professionalism of the investor, a single standardized solution can be offered at this point, which adheres to both conditions.

5.8 Summary

Robo-advisory thrives on automation and quant models enable a high degree of automation on the strategy side. While quant models can work well in more conventional environments (this will remain the preferred choice for institutional investors for the foreseeable future), robo-advisory and quantitative investment strategies represent a very good structural fit. The cost savings that robo-advisory offers compared to conventional asset management can largely be passed on to the investor. This can deliver added value, especially compared to the highly standardized solutions for small investors that are otherwise customary on the market. Within the now broad spectrum of quantitative strategies, however, a distinction must be made: time-stable, low-maintenance models are preferable, which implicitly amounts to a renunciation of return forecasts. Strategies from the field of postmodern portfolio theory are optimal in this respect, which focus specifically on risk budgeting and/or risk minimization.

The strength of the models described here lies, among other things, in the control of portfolio risks not previously achieved by less affluent investors, right up to the specification of maximum loss limits or the specification of an investment horizon, according to which it is highly probable that the invested capital will be available again at least nominally. The models can also be set up in such a way that they are customizable by the investor within the previously defined framework. The complexity at the level of the selection process must accommodate the investor's level of experience. Less experienced investors should therefore continue to have access to a manageable number of standardized solutions in the future. While such standardized solutions are still often the state of play for all investment groups, the approaches shown here can also appeal to more demanding investors.

References

Bankenverband. 2017. *Positionspapier des Bankenverbandes zu Robo-Advice*. Tech. rep. https://bankenverband.de/media/files/2017_03_20Positionspapier_RoboAdvice.pdf.

William A. Brock, Josef Lakonishok, and Blake LeBaron. 1992. "Simple Technical Trading Rules and the Stochastic Properties of Stock Returns". *Journal of Finance* 47(5): 1731–1764. https://doi.org/10.1111/j1540-6261.1992.tb04681.x.

Yves Choueifaty and Yves Coignard. 2008. "Toward Maximum Diversification". *The Journal of Portfolio Management* 35(1): 40–51. https://doi.org/10.3905/JPM.2008.35.1.40.

Yves Choueifaty, Tristan Froidure, and Julien Reynier. 2013. "Properties of the Most Diversified Portfolio". *The Journal of Investment Strategies* 2: 49–70. https://doi.org/10.21314/JOIS.2013.033.

Roger G. Clarke, Harindra de Silva, and Steven R. Thorley. 2011. "Minimum Variance Portfolio Composition". *The Journal of Portfolio Management* 37(2): 31–45. https://doi.org/10.2139/ssrn.1549949.

Roger G. Clarke, Harindra de Silva, and Steven R. Thorley. 2013. "Risk Parity, Maximum Diversification, and Minimum Variance: An Analytic Perspective". *The Journal of Portfolio Management* 39(3): 39–53. https://doi.org/10.2139/ssrn.1977577.

Jon Danielsson and Jean-Pierre Zigrand. 2006. "On Time-Scaling of Risk and the Square–Root–of–Time Rule". *Journal of Banking & Finance* 30(10): 2701–2713. https://doi.org/10.1016/j.jbankfin.2005.10.002.

Richard C. Grinold and Ronald N. Kahn. 2012. *Active Portfolio Management.* McGraw-Hill Professional.

Jonathan W. Lam. 2016. *Robo-Advisors: A Portfolio Management Perspective.* New Haven, CT: Yale College. https://economics.yale.edu/sites/default/files/files/Undergraduate/Nominated%20Senior%20Essays/2015-16/Jonathan_Lam_Senior%20Essay%20Revised.pdf.

Bob Litterman. 2003. *Modern Investment Management: An Equilibrium Approach.* Hoboken: Jon Wiley & Sons.

Harry M. Markowitz. 1952. "Portfolio Selection". *Journal of Finance* 7(1): 77–91.

Attilio Meucci. 2009. *Risk and Asset Allocation.* Springer Finance.

Thorsten Rühl. 2001. "Studien zur Verbesserten Ausnutzung des Informationsgehaltes von Multisensorsystemen". PhD thesis. Institut für Angewandte Physik Justus-Liebig-Universität Gießen. http://geb.uni-giessen.de/geb/volltexte/2002/709/pdf/d020009.pdf.

Stephen Satchell. 2003. In *Advances in Portfolio Construction and Implementation (Quantitative Finance)*, ed. Alan Scowcroft. Butterworth-Heinemannn.

Donald F Specht. 1991. "A General Regression Neural Network". *IEEE Transactions on Neural Networks* 2(6): 568–576. https://doi.org/10.1109/72.97934.

Jérôme Teiletche, Thierry Roncalli, and Sébastien Maillard. 2010. "The Properties of Equally Weighted Risk Contribution Portfolios". *Journal of Portfolio Management* 36(4): 60–70. https://doi.org/10.3905/jpm.2010.36.4.06.

CHAPTER 6

Analysis of the Use of Robo-Advisors as a Replacement for Personal Selling

Goetz Greve and Frederike Meyer

Contents

G. Greve (✉)
Hamburg School of Business Administration (HSBA), Hamburg, Germany
e-mail: goetz.greve@hsba.de

F. Meyer
Sparkasse Harburg-Buxtehude, Hamburg, Germany
e-mail: frederike.meyer@spkhb.de

© The Author(s) 2021

P. Scholz (ed.), *Robo-Advisory*, Palgrave Studies in Financial Services Technology, https://doi.org/10.1007/978-3-030-40818-3_6

6.1 Introduction

With the development of robo-advisors and their adoption by consumers, the sales management of financial services is changing rapidly. This development may lead to the disintermediation of salespeople, as technologies emancipate customers and they can inform themselves about offerings. Consequently, customers may not view the buying process as one that is necessarily driven by humans. Whereas research has already surveyed the perspective of salesperson technology adoption (e.g. Ahearne and Rapp 2010; Verma et al. 2016; Moncrief 2017), little is known about the customer perspective when it comes to customer–salesperson interaction technologies. The goal of this study, therefore, is to compare robo-advisors with salespersons and hybrid solutions and their impact on behavioral constructs. Using experiments, we contribute to the literature by investigating how different forms of customer–salesperson interaction technologies impact customer perception with respect to trust, perceived risk, psychological reactance, and perceived use. Second, we contribute to the growing robo-advisory literature by demonstrating how robo-advisory services are perceived by customers.

6.2 Robo-Advisors as Customer–Salesperson Interaction Technologies

Robo-advisors can be defined as automatic, web-based tools that provide customers financial advice without human intervention using computer algorithms to manage client portfolios. Robo-advisors are seen as a true asset management innovation with significant growth potential. Accordingly, fintech start-ups as well as traditional banks and asset managers are increasingly showing interest in offering robo-advisory services as these may offer the possibility to replace human salesperson advice by robo-advice (Wirtz et al. 2018). Traditionally, the salesperson has been the leading player in a buyer–seller dyad (Zboja et al. 2016). However, driven by the development of digital technologies and tools, consumers have changed their buying behavior in accordance with the adoption of interactive new media (Crittenden et al. 2010) as information about products and services is available online without any salesperson contact. As interactive new media increases the availability of online information to a maximum, it is not surprising that, on the one hand, rising customer expectations and customer avoidance of personal buyer–seller negotiations on the other impact personal selling and sales management (Anderson 1996). Today, for a large variety of products and services, consumers are more sophisticated than ever before, and are buying without the aid of sales personnel (Verma et al. 2016). Ahearne and Rapp (2010) conclude that the more upcoming technologies enable the consumer to make an educated buying decision on his/her own, the higher is the probability that disintermediation of salespeople will occur. On a salesperson–customer interface technology continuum, they propose on the one end salesperson-specific technologies, that is, technologies used solely by the sales representatives. They point out that in most circumstances customers do not see or are not even aware of technology use. On the other end, they consider customer-specific technologies as technologies used solely by the customer. They propose that these technologies may eliminate the role of the salesperson. This end, therefore, reflects a potential for disintermediation of salespeople. It can be proposed that consumers may no longer be exclusively driven by human contact or they even may not require human contact for making buying decisions anymore (Moncrief 2017; Ahearne and Rapp 2010; Moore 2015; Marshall et al. 2012). This argumentation may be true for a large variety of existing products that are already sold via e-commerce solutions. However, it can be questioned whether this proposition also holds true for complex financial products. With respect to technology, the research on sales from the salesperson's perspective has been extensive so far (Sharma et al. 2010). Studies have revealed that for consumer self-service technologies, acceptance is driven by ability and the ease of use (Meuter et al. 2005; Brown et al. 2008). However, we see a distinct gap in the analysis of different levels of customer–salesperson interaction technologies. So far, it remains unknown whether customer-specific interaction technologies (CSIT) impact the disintermediation of salespeople, especially within the field of complex financial products. Consequently, we follow the argumentation of Anderson (1996) and Ahearne and Rapp (2010) and test the effects of either robo-advisors as a form of selling without human contact or human-driven personal selling in an investment-advisory setting. As a third option, we consider a hybrid form of personal selling with technology support, thereby depicting different positions on the proposed salesperson–customer interface technology continuum.

6.3 Theoretical Background and Research Propositions

As suggested by the relationship marketing paradigm as well as social exchange theory, trust (Pavlou 2003), perceived risk (Sheth et al. 1999; Pavlou 2003), psychological reactance (Brehm 1968), and perceived use (Thibaut and Kelley 1959) are important concepts of buyer–seller interactions-especially in the online (e-commerce) context. In accordance with the conceptualization of Ahearne and Rapp (2010), we consider personal selling as close relational exchanges with a maximum of social interaction, thereby describing the maximum of salesperson-oriented interaction technologies on the proposed salesperson–customer interface technology continuum. On the other end of the continuum, we consider robo-advisors as discrete exchanges characterized by nil social interaction. Robo-advisors are solely used by the customers, and, so, eliminate the role of the salesperson. Robo-advice can be defined as digital investment advice tools that match consumers on the basis of their personal preference to financial products (Ringe and Ruof 2018). It should be noted that a key characteristic of robo-advisors is the absence of any human contact between the advisor and the customer (Fisch et al. 2017). Hence, employing these concepts in the uncertain context of robo-advisors as a new form of online selling is also reasonable. Therefore, perceptions of trust, risk, psychological reactance, and use are likely to be important factors in consumer acceptance of robo-advisors.

6.3.1 Trust

According to social exchange theory (Thibaut and Kelley 1959), people form exchange relationships on the basis of trust and perceived risk. Trust is defined as "confidence in the exchange partner's reliability and integrity" (Morgan and Hunt 1994). It is viewed as an important component for successful relationship building (Stewart and Pavlou 2002). However, exchange relationships that are likely to cost more than the potential reward will be avoided. On the internet, customers typically perceive higher risk compared to a conventional shopping environment (Tan 1999) due to distance, virtual identity, and lack of regulation. Therefore, trust is the preliminary condition for consumers' e-commerce participation. In the context of robo-advisors, it becomes clear that there is a risk of monetary loss since consumers have to solely rely on online information and, so, may become more vulnerable to inaccurate, incomplete, or wrong information provided by robo-advisors. In addition, there is a risk of privacy loss by providing personal information to robo-advisors. Consequently, the importance of trust is elevated in e-commerce because of a possible high degree of uncertainty and risk present in most online transactions (Jarvenpaa et al. 1999). Hence, we propose Hypothesis 1: A higher level of CSIT will lead to a lower level of trust.

6.3.2 *Perceived Risk*

Perceived risk is defined as the consumer's subjective belief of suffering a loss in the pursuit of a desired outcome (Bauer 1960; Sheth and Parvatiyar 1995). Within the context of e-commerce, perceived risk is regarded as an important driver of consumer intentions to buy. In the online context, the distant and impersonal nature of the online environment has been associated with environmental uncertainty (e.g. technology) or behavioral uncertainty (e.g. relational) (Bensaou and Venkatraman 1996). Environmental uncertainty may arise out of the missing control of the consumer as regards its information. Although the seller has an important influence on the security of the transaction medium (e.g. encryption, authentication, and firewalls), there is still a possibility of third parties compromising the transaction process. Behavioral uncertainty may arise because of opportunistic behavior of the seller, including false information, misleading product presentations, and misleading advertising. Perceived risk has been shown to negatively influence consumer online buying intentions (Jarvenpaa et al. 1999; Pavlou 2003). The perceived risk associated with robo-advisors may reduce perceptions of behavioral and environmental control, and this lack of control may negatively influence buying intentions. Within the context of robo-advisors, we assume that customer-specific technologies will result in a higher level of customer's perceived risk due to technology-driven environmental uncertainty. Therefore, we derive Hypothesis 2: A higher level of CSIT will lead to a higher level of perceived risk.

6.3.3 *Psychological Reactance*

The socio-psychological theory of psychological reactance indicates that when a perceived freedom is eliminated or threatened with elimination, the individual will be motivated to reestablish that freedom (Brehm 1968). Relationship marketing is widely based on commitment (Morgan and Hunt 1994). This commitment can be either formalized by a contractual setting or not. However, formalization may be interpreted by the customer as a thread to its freedom of choice as for promotional influence or manipulative advertisement (Clee and Wicklung 1980). Consequently, customers will show psychological reactance. Transferred to the context of CSIT, consumers are confronted with many unwanted marketing communications through various channels. In addition, they may recognize persuasive tactics of salespersons and their attempt to push the purchase of a specific product. These forced intrusions are perceived as threats to their freedom of choice (Martin and Murphy 2017). Consequently, consumers may show psychological reactance, which, in turn, may motivate consumers to regain their lost freedom (Edwards et al. 2002). We, therefore, assume that a higher level of CSIT will lead to a lower level of psychological reactance. Hence, we state Hypothesis 3: A higher level of CSIT will lead to a lower level of physiological reactance.

6.3.4 Perceived Use

According to social exchange theory, "perceived use" is a key construct to explain why consumers continue or complete a social interaction. Thibaut and Kelley (1959) postulate that a consumer judges the use of an interaction on the basis of a comparison level describing an individual cost-benefit ratio. Applied to technology, the term "perceived usefulness" is defined as the individual's perception that the use of the new technology will enhance or improve his or her performance (Davis 1989). In the context of robo-advisors, usefulness refers to the degree to which consumers believe the use of robo-advisors will improve their performance or productivity, thereby enhancing the outcome of investment advice. However, in comparison with investment advice by salespeople, the relationship between robo-advisors and perceived usefulness is not clear. With decreasing financial literacy in society (Hastings et al. 2013), we propose an opposite trend toward salespeople interaction in investment advice. Hence, we propose Hypothesis 4: A higher level of CSIT will lead to a lower level of perceived use.

6.4 Experimental Design and Data Collection

To test the derived framework, a single factorial between-subjects experiment was conducted with the experimental factors personal salesperson advice, robo-advice, and hybrid advice (robo-advice + salesperson advice). For the experiment, three treatment groups of randomly selected customers of a German savings bank were formed, and each of them was exposed to a different stimulus. For the purpose of this research, it was decided to use real robo-advisors and a real investment advice setting at a savings bank, including real salespeople and robo-advisors, in order to have a high degree of realism (Geuens and De Pelsmacker 2017). Any other influences, such as the local conditions, were reduced, or, at least, homogenized, for all groups to limit potential biases. The manipulated stimuli as well as the entire study design were pretested among participants. In toto, a convenience sample of 75 participants was collected and randomly assigned to the three treatment groups, resulting in three independent samples, with 25 participants each using a parallel design. The participants were aged between 18 and 35 years. Furthermore, the sample was equally distributed between women and men. To measure the surveyed constructs, multi-items were used that were already established by prior research and are accepted in literature; most items were then measured on a seven-point Likert scale. The measurement of trust was done by applying the scale of Lee and Henderson (1992). Perceived risk was measured by adapting the scale of Sheth and Parvatiyar (1995). Psychological reactance was measured by adapting the scale of Hong and Page (1989). Perceived use was measured by adapting the scale of Zaichkowsky (1994).

6.5 Results

6.5.1 *Data Preparation and Manipulation Checks*

Almost all constructs had alpha scores well above 0.7 (trust = 0.911; perceived risk = 0.785; psychological reactance = 0.686; perceived use = 0.889), displaying a good level of reliability (Nunnally and Bernstein 1994). Pearson's Chi-square test and Kruskal–Wallis H-test were applied to check for structural equality based on gender and age. The results indicate no significant difference between the three experimental groups. Correspondingly, the groups can be compared and used to reliably and validly investigate the proposed differences in consumer behavioral constructs under different conditions. Statistical measures show that the manipulation of salesperson advice, robo-advice, as well as hybrid advice was successful. To check the manipulations of salesperson advice (salesperson advice, robo-advice, and hybrid advice), pairwise Kruskal–Wallis tests were applied, resulting in Chi-square values ranging from 8.967 (psychological reactance) to 30.276 (trust) on a <0.01 significance level.

6.5.2 *Hypotheses Test*

To test our propositions, we conducted planned comparisons. Kruskal–Wallis H-tests for each treatment group and constructs showed that there are statistically significant differences in the dependent variables between the different advice treatments. Table 6.1 shows the respective results. For all surveyed constructs, we can partially confirm our hypotheses. The comparison for robo-advice between groups indeed revealed that these forms of customer-specific interaction technology achieve the highest ranks for perceived risk and psychological reactance and the lowest ranks for trust

Table 6.1 Results of Kruskal–Wallis H tests

Construct	Hyp.	*N*	*Alpha*[a]	Mean	SD	χ^2	*p*	Rank
Trust	H1	75	0.911	5.71	1.11	30.288		
Robo-advice		25					0.00	18.66
Hybrid advice		25					0.00	49.98
Salesperson advice		25					0.00	45.36
Perceived risk	H2	75	0.785	5.62	0.91	22.398		
Robo-advice		25					0.00	54.78
Hybrid advice		25					0.00	29.52
Salesperson advice		25					0.00	29.70
Psychological	H3		0.686	2.07	0.84	8.967		
Robo-advice		25					0.01	48.24
Hybrid advice		25					0.01	30.54
Salesperson advice		25					0.01	35.22
Perceived use	H4		0.889	5.41	1.21	22.340		
Robo-advice		25					0.00	21.30
Hybrid advice		25					0.00	47.84
Salesperson advice		25					0.00	44.86

[a] Cronbach's alpha scores

and perceived use compared to the other groups. However, comparing salesperson advice and hybrid advice, interestingly, hybrid advice showed higher ranks than salesperson advice for trust and perceived use.

For risk and psychological reactance, the groups indicated lower ranks compared to salesperson advice. By this, we can conclude that our findings are somewhat contradictory to the conceptualization of Ahearne and Rapp (2010). We cannot confirm a linear increase of the measured constructs with respect to salesperson–customer interaction technologies. Interestingly, the hybrid advice seems to outperform the other two forms of interaction, emphasizing the point that the human touch in selling still seems to matter. This might be explained by the fact that in the case of complex financial products, consumers still prefer a human touch. This can be concluded by the fact that the hybrid advice (robo-advice + salesperson advice) scores higher for all researched constructs. Hence, personal selling for complex financial products can still be the key, even in the presence of robo-advisors. However, it can be assumed that the detected phenomenon may not necessarily exist forever. It can be expected that, on the one hand, robo-advisors will develop their capabilities further on with the help of artificial intelligence. On the other hand, it can be proposed that consumers will adapt to the fact that more and more robots will help them with their buying decision-making, resulting in a higher acceptance rate of robo-advisors, and also for the financial services industry.

6.6 Managerial Implications and Limitations

Our findings have important implications for the implementation of robo-advisors in financial services. As the momentary trend of shifting resources from salesperson-centric technologies to customer-centric technologies is still growing strong, more and more companies evaluate the implementation of, for example, robo-advisors as a form of CSIT without any human touch. However, our findings suggest that companies should not underestimate the value of personal selling from a consumer's perspective. Hence, based on our experimental setting, we can only partially agree to the conceptualization of Ahearne and Rapp (2010), who suggest disintermediation of salespeople by machines. Our results rather indicate that banking customers still favor a hybrid of personal selling and robo-advisory services. This result is possibly linked to the fact that assessment advisory has a high degree of complexity for the consumer. Based on this fact, we assume that a disintermediation of the salesperson does not occur in a strong way as long as the financial product has a high degree of complexity, specialization, and individualization. Financial products with a lower degree of complexity have a higher risk of the disintermediation of the salesperson, which can already be seen in the area of banking accounts (e.g. Paypal). However, this must not necessarily be the case in the future. The further development of artificial intelligence and, therefore, the improvement of robo-advisors may increase the risk for traditional banking services like personal investment advisors to be disintermediated by robo-advisors. Regarding the possible limitations of the study, it needs to be said that the external validity of the study may be limited, as the data collection was made using an experimental design. Hence, the relationship between CSIT and the surveyed constructs needs to be examined in real-life settings.

References

Ahearne, Michael, and Adam Rapp. 2010. "The Role of Technology at the Interface Between Salespeople and Consumers." *Journal of Personal Selling and Sales Management* 30 (2): 111–120. https://doi.org/10.2753/PSS0885-3134300202.

Anderson, Rolph E. 1996. "Personal Selling and Sales Management in the New Millennium." *Journal of Personal Selling and Sales Management* 16 (4): 17–32.

Bauer, Raymond A. 1960. "Consumer Behavior as Risk Taking." In *Dynamic Marketing for a Changing World*, ed. R.S. Hancock, 389–398. Chicago: American Marketing Association.

Bensaou, Ben M., and Venkat N. Venkatraman. 1996. "Inter-Organizational Relationships and Information Technology: A Conceptual Synthesis and a Research Framework." *European Journal of Information Systems* 5: 84–91.

Brehm, Jack. 1968. "A Theory of Psychological Reactance." *The American Journal of Psychology* 81. https://doi.org/10.2307/1420824.

Brown, Susan A., Viswanath Venkatesh, Jason Kuruzovich, and Anne P. Massey. 2008. "Expectation Confirmation: An Examination of Three Competing Models." *Organizational Behavior and Human Decision Processes* 105 (1): 52–66. https://doi.org/10.1016/j.obhdp.2006.09.008.

Clee, Mona A., and Robert A. Wicklund. 1980. "Consumer Behavior and Psychological Reactance." *Journal of Consumer Research* 6 (4): 389–405. https://doi.org/10.1086/208782.

Crittenden, Victoria, Robert Peterson, and Gerald Albaum. 2010. "Technology and Business-to-Consumer Selling: Contemplating Research and Practice." *Journal of Personal Selling and Sales Management* 30: 103–110. https://doi.org/10.2753/PSS0885-3134300201.

Davis, Fred D. 1989. "Perceived Usefulness, Perceived Ease of Use, and User Acceptance of Information Technology." *MIS Quarterly* 13 (3): 319–340.

Edwards, Steven, Hairong Li, and Joo-Hyun Lee. 2002. "Forced Exposure and Psychological Reactance: Antecedents and Consequences of the Perceived Intrusiveness of Pop-Up Ads." *Journal of Advertising* 31 (3): 83–95. https://doi.org/10.1080/00913367.2002.10673678.

Fisch, Jill E., Marion Laboure, and John A. Turner. 2017. *The Economics of Complex Decision Making: The Emergence of the Robo Adviser.* Last Accessed: 25 November 2019. http://www.geog.ox.ac.uk/events/170911/Robo-vs-Human-Advisers-Aug-28.pdf.

Geuens, Maggie, and Patrick De Pelsmacker. 2017. "Planning and Conducting Experimental Advertising Research and Questionnaire Design." *Journal of Advertising* 46 (1): 83–100. https://doi.org/10.1080/00913367.2016.1225233.

Hastings, Justine S., Brigitte C. Madrian, and William L. Skimmyhorn. 2013. "Financial Literacy Financial Education and Economic Outcomes." *The Annual Review of Economics* 5: 347–373.

Hong, Sung-Mook, and Sandra Page. 1989. "A Psychological Reactance Scale: Development, Factor Structure and Reliability." *Psychological Reports* 64 (3): 1323–1326. https://doi.org/10.2466/pr0.1989.64.3c.1323.

Jarvenpaa, Sirkka L., Noam Tractinsky, and Michael R. Vitale. 1999. "Consumer Trust in an Internet Store." *Information Technology and Management* 1: 45–71.

Lee, Soonchul, and John C. Henderson. 1992. "Managing I/S Design Teams: A Control Theories Perspective." *Management Science* 38 (6): 757–777.

Marshall, Greg, William Moncrief, John Rudd, and Nick Lee. 2012. "Revolution in Sales: The Impact of Social Media and Related Technology on the Selling Environment." *Journal of Personal Selling and Sales Management* 32: 349–363. https://doi.org/10.2307/23483286.

Martin, Kelly D., and Patrick E. Murphy. 2017. "The Role of Data Privacy in Marketing." *Journal of the Academy of Marketing Science* 45 (2): 135–155. https://doi.org/10.1007/s11747-016-0495-4.

Meuter, Matthew, Mary Bitner, Amy Ostrom, and Stephen Brown. 2005. "Choosing Among Alternative Service Delivery Modes: An Investigation of Customer Trial of Self-Service Technologies." *Journal of Marketing* 69 (2): 61–83. https://doi.org/10.1509/jmkg.69.2.61.60759.

Moncrief, William. 2017. "Are sales As We Know It Dying ... or Merely Transforming?" *Journal of Personal Selling and Sales Management* 37: 1–9. https://doi.org/10.1080/08853134.2017.1386110.

Moore, Sarah G. 2015. "Attitude Predictability and Helpfulness in Online Reviews: The Role of Explained Actions and Reactions." *Journal of Consumer Research* 42 (1): 30–34. https://doi.org/10.1093/jcr/ucv003.

Morgan, Robert M., and Shelby D. Hunt. 1994. "The Commitment-Trust Theory of Relationship Marketing." *Journal of Marketing* 58 (3): 20–38.

Nunnally, Jum C., and Ira H. Bernstein. 1994. *Psychometric Theory*. 3rd ed. New York: McGraw-Hill.

Pavlou, Paul A. 2003. "Consumer Acceptance of Electronic Commerce: Integrating Trust and Risk with the Technology Acceptance Model." *International Journal of Electronic Commerce* 7: 101–134.

Ringe, Wolf-Georg, and Christopher Ruof. 2018. *A Regulatory Sandbox for Robo Advice*. eng. ILE Working Paper Series 14. Hamburg. http://hdl.handle.net/10419/179514.

Sharma, Dheeraj, Jule Gassenheimer, and Bruce Alford. 2010. "Internet Channel and Cannibalization: An Empirical Assessment of Sales Agents' Perspective." *Journal of Personal Selling and Sales Management* 30: 209–222. https://doi.org/10.2753/PSS0885-3134300302.

Sheth, Jagdish N., and Atul Parvatiyar. 1995. "The Evolution of Relationship Mar keting." *International Business Review* 4 (4): 397–418.

Sheth, Jagdish N., Banwari Mittal, and Bruce Newman. 1999. *Customer Behavior: Consumer Behavior and Beyond*. Cincinnati: South-Western College Publishing.

Stewart, David W., and Paul A. Pavlou. 2002. "From Consumer Response to Active Consumer: Measuring the Effectiveness of Interactive Media." *Journal of the Academy of Marketing Science* 30 (4): 376–396. https://doi.org/10.1177/009207002236912.

Tan, Soo J. 1999. "Strategies for Reducing Consumers' Risk Aversion in Internet Shopping." *Journal of Consumer Marketing* 16: 163–180.

Thibaut, John W., and Harold H. Kelley. 1959. *The Social Psychology of Groups*. New York: Wiley.

Verma, Varsha, Dheeraj Sharma, and Jagdish N. Sheth. 2016. "Does Relationship Marketing Matter in Online Retailing? A MetaAnalytic Approach." *Journal of the Academy of Marketing Science* 44: 206–217.

Wirtz, Jochen, Paul Patterson, Werner Kunz, Thorsten Gruber, Vinh Lu, Stefanie Paluch, and Antje Martins. 2018. "Brave New World: Service Robots in the Frontline." *Journal of Service Management* 29 (5): 907–931. https://doi.org/10.1108/JOSM-04-2018-0119.

Zaichkowsky, Judith L. 1994. "The Personal Involvement Inventory: Reduction, Revision, and Application to Advertising." *Journal of Advertising* 23 (4): 59–70. https://doi.org/10.1080/00913367.1943.10673459.

Zboja, James J., Ronald A. Clark, and Diana L. Haytko. 2016. "An Offer You Can't Refuse: Consumer Perceptions of Sales Pressure." *Journal of the Academy of Marketing Science* 44 (6): 806–821. https://doi.org/10.1007/s11747-015-0468-z.

CHAPTER 7

Regulation of Robo-Advisers in the United States

Melanie L. Fein

Contents

This chapter was written as of July 2019. It does not reflect any changes in the law that might have occurred or may occur after then.

M. L. Fein (✉)
Fein Law Offices, Great Falls, VA, USA
e-mail: melanie@feinlawoffices.com

© The Author(s) 2021

P. Scholz (ed.), *Robo-Advisory*, Palgrave Studies in Financial Services Technology, https://doi.org/10.1007/978-3-030-40818-3_7

7.1 Introduction

Robo-advisers operate in the United States under a regulatory framework that has evolved over decades to govern the securities industry. This framework was not designed with robo-advisers in mind but has been applied by regulators to robo-adviser activities. US regulators have not adopted regulations specifically aimed at robo-advisers but rather have applied the existing framework to them.

7.2 Investment Advisers Act of 1940

The Investment Advisers Act of 1940 (Advisers Act) is the federal statute governing investment advisers in the United States. Robo-advisers are treated as investment advisers under this act, which is administered by the Securities and Exchange Commission (SEC). The SEC has said that "robo-advisers are subject to all of the requirements of the Advisers Act."[1]

The Advisers Act does not provide a comprehensive regulatory regime but rather provides a principles-based overlay for investment adviser activities. As the SEC has explained:

> Unlike the laws of many other countries, the US federal securities laws do not prescribe minimum experience or qualification requirements for persons providing investment advice. They do not establish maximum fees that advisers may charge. Nor do they preclude advisers from having substantial conflicts of interest that might adversely affect the objectivity of the advice they provide. Rather, investors have the responsibility, based on disclosure they receive, for selecting their own advisers, negotiating their own fee arrangements, and evaluating their advisers' conflicts.[2]

7.2.1 Registration of Robo-Advisers

Section 203 of the Advisers Act provides that "it shall be unlawful for any investment adviser, unless registered under this section, to make use of the mails or any means or instrumentality of interstate commerce in

[1] See Division of Investment Management (2017) ("SEC Staff Guidance"), n.8. Because the SEC uses the spelling "robo-advisers" (as opposed to "robo-advisor"), this chapter will use the SEC's spelling for consistency.

[2] Investment Advisers Act of 1940, 17 C.F.R. § 275 and 279 (2008).

connection with his or its business as an investment adviser." Robo-advisers come within the definition of "investment adviser" in the Advisers Act and thus generally cannot conduct business without registering under the Act. The Act defines "investment adviser" to mean:

> any person who, for compensation, engages in the business of advising others, either directly or through publications or writings, as to the value of securities or as to the advisability of investing in, purchasing, or selling securities, or who, for compensation and as part of a regular business, issues or promulgates analyses or reports concerning securities ...[3]

Advisers that meet this definition and have assets under the management of $100 million or more are required to register with the SEC as investment advisers.[4] Advisers with less than $100 million in assets under management must register with a state regulator rather than the SEC. State statutes generally require advisers to register in every state in which the adviser obtains more than a de minimis number of clients.

Under an SEC rule, investment advisers that provide advice through interactive websites, such as robo-advisers, may register with the SEC rather than the states.[5] The rule is intended to lessen the burden of multi-state registration on investment advisers that operate exclusively through the Internet.[6]

[3] Investment Advisers Act of 1940, Pub. L. No. 115–417 § 202(a)(11) (2019). State securities laws similarly define "investment adviser".

[4] There are currently over 8000 SEC-registered investment advisers in the United States. A non-US adviser giving advice to US persons must register with the SEC unless an exemption from registration is available (in which case it still may be subject to state registration). Advisers that would otherwise be obligated to register with 15 or more states may register with the SEC.

[5] SEC Rule 203A-2(e) defines an "interactive website" as a website in which computer software-based models or applications provide investment advice to clients based on personal information provided by each client through the website. An adviser relying on the exemption may not use its advisory personnel to elaborate or expand on the investment advice provided by its interactive website, or otherwise provide investment advice to its Internet clients, except as permitted by the rule's de minimis exception.

[6] Investment Advisers Act of 1940, 17 C.F.R. § 275.203A-2(e) (2002).

7.2.2 *Anti-fraud Provisions*

Section 206 of the Advisers Act contains the Act's anti-fraud provisions, which form the basis of the SEC's principles-based regulation of investment advisers. These provisions provide as follows:

> It shall be unlawful for any investment adviser by the use of mails or any means or instrumentality of interstate commerce, directly or indirectly—
>
> 1. to employ any device, scheme, or artifice to defraud any client or prospective client;
> 2. to engage in any transaction, practice, or course of business which operates as a fraud or deceit upon any client or prospective client;
> 3. to act as principal for his own account, knowingly to sell any security to or purchase any security from a client, or acting as broker for a person other than such client, knowingly to effect any sale or purchase of any security for the account of such client, without disclosing to such client in writing before the completion of such transaction the capacity in which he is acting and obtaining the consent of the client to such transaction; or
> 4. to engage in any act, practice, or course of business which is fraudulent, deceptive, or manipulative[7]

These provisions are intended to prevent investment advisers from defrauding, deceiving, or manipulating their clients. They do not by their terms impose a fiduciary standard of care on an adviser's investment recommendations or require an adviser to act in the best interest of its clients. However, as discussed below, these provisions have been interpreted by the courts and the SEC as imposing a fiduciary duty on investment advisers.

7.2.3 *Fiduciary Duty*

Based on the Adviser Act's anti-fraud provisions, the courts and the SEC have stated that investment advisers are "fiduciaries." [8] However, the SEC

[7] Investment Advisers Act of 1940, 15 U.S.C. §80b-6 (2020). The act authorizes the SEC to adopt rules and grant exemptions from these prohibitions.

[8] See SEC v. Capital Gains Research Bureau, Inc., 375 U.S. 180, 194 (1963). Citing that case, the SEC has stated, "Under federal law, an investment adviser is a fiduciary." See Securities and Exchange Commission (2019), at 2.

has not by regulation addressed what the fiduciary duties of investment advisers are. Rather, the SEC historically has taken a "principles-based" approach whereby investment advisers are expected to act in their customers' best interest in accordance with common law fiduciary principles. The SEC recently reaffirmed its principles-based approach in an official interpretation of the fiduciary duty of investment advisers.[9] The SEC's "Fiduciary Interpretation" states:

> The Advisers Act establishes a federal fiduciary duty for investment advisers. This fiduciary duty is based on equitable common law principles and is fundamental to advisers' relationships with their clients under the Advisers Act. The investment adviser's fiduciary duty is broad and applies to the entire adviser-client relationship.[10]

An adviser's fiduciary duty is comprised of a duty of care and a duty of loyalty:

> The fiduciary duty an investment adviser owes to its client under the Advisers Act [...] comprises a duty of care and a duty of loyalty...[11]
> This fiduciary duty requires an adviser to adopt the principal's goals, objectives, or ends. This means the adviser must, at all times, serve the best interest of its client and not subordinate its client's interest to its own. In other words, the investment adviser cannot place its own interests ahead of the interests of its client. This combination of care and loyalty obligations has been characterized as requiring the investment adviser to act in the "best interest" of its client at all times. In our view, an investment adviser's obligation to act in the best interest of its client is an overarching principle that encompasses both the duty of care and the duty of loyalty.[12]

[9] See Securities and Exchange Commission (2019) (the "Fiduciary Interpretation").

[10] SEC, *Fiduciary Interpretation*, 6–8. The SEC acknowledged that the fiduciary duty is not specifically defined in the Advisers Act or in SEC rules but reflects a Congressional recognition "of the delicate fiduciary nature of an investment advisory relationship" as well as a Congressional intent to "eliminate, or at least to expose, all conflicts of interest which might incline an investment adviser – consciously or unconsciously – to render advice which was not disinterested."

[11] SEC, *Fiduciary Interpretation*, 2.

[12] SEC, *Fiduciary Interpretation*, 7.

The SEC emphasized that the duty to provide investment advice that is in the "best interest" of the client includes a duty to provide advice that is "suitable" for the client.[13] In order to provide such advice, the SEC said an adviser must have a reasonable understanding of the client's objectives. For retail clients, the SEC said the basis for such a reasonable understanding generally would include an understanding of the investment profile.[14] How an adviser develops a reasonable understanding "will vary based on the specific facts and circumstances, including the nature of the client, the scope of the adviser-client relationship, and the nature and complexity of the anticipated investment advice."[15] The SEC stated:

> In order to develop a reasonable understanding of a retail client's objectives, an adviser should, at a minimum, make a reasonable inquiry into the client's financial situation, level of financial sophistication, investment experience, and financial goals (which we refer to collectively as the retail client's "investment profile"). For example, an adviser undertaking to formulate a comprehensive financial plan for a retail client would generally need to obtain a range of personal and financial information about the client such as current income, investments, assets and debts, marital status, tax status, insurance policies, and financial goals.[16]

This formulation allows flexibility for robo-advisers to comply with the fiduciary duty to provide suitable investment advice, provided the robo-adviser also complies with the duty of loyalty. Under the duty of loyalty, an investment adviser must "eliminate or make full and fair disclosure of all conflicts of interest which might incline an investment adviser-consciously or unconsciously-to render advice which is not disinterested, such that a client can provide informed consent to the conflict."[17]

The Fiduciary Interpretation specifically recognizes robo-advisers as investment advisers, stating that robo-advisers are subject to the fiduciary obligations that apply to investment advisers under the Investment Advisers Act of 1940:

[13] SEC, *Fiduciary Interpretation*, 12.
[14] SEC, *Fiduciary Interpretation*, 13.
[15] SEC, *Fiduciary Interpretation*, 13.
[16] SEC, *Fiduciary Interpretation*, 12–13.
[17] SEC, *Fiduciary Interpretation*, 12–13.

> This [Fiduciary] Interpretation also applies to automated advisers, which are often colloquially referred to as "robo-advisers." Automated advisers, like all SEC-registered investment advisers, are subject to all of the requirements of the Advisers Act, including the requirement that they provide advice consistent with the fiduciary duty they owe to their clients.[18]

The SEC emphasized that the fiduciary duty "follows the contours of the relationship between the adviser and its client, and the adviser and its client may shape that relationship by agreement ... we recognize that investment advisers provide a wide range of services ... and serve a large variety of clients."[19] The SEC noted that the fiduciary duty provides flexibility for different business models:

> In our experience, the principles-based fiduciary duty imposed by the Advisers Act has provided sufficient flexibility to serve as an effective standard of conduct for investment advisers, regardless of the services they provide or the types of clients they serve.[20]

Although the SEC did not specifically endorse the robo-adviser business model, the SEC's statements implicitly recognize that the fiduciary duty under the Advisers Act can accommodate robo-advisers as well.

Depending on how the "best interest" standard is interpreted and applied, however, it could be problematic for some robo-advisers. For example, the best interest standard can be read to require a robo-adviser to provide personalized advice in the best interest of each customer individually rather than its customers collectively. Moreover, it obligates the robo-adviser, not the customer, to make an affirmative determination that an investment recommendation is in the best interest of that customer. It is not clear that all robo-advisers can meet this standard.

The Interpretation requires an investment adviser to provide investment advice in the best interest of its client "based on the client's objectives."[21] Thus, the Interpretation appears to endorse "goal-based" investment advice that robo-advisers are expected to offer. At the same time, however, the Interpretation would seem to require that algorithms used by robo-

[18] SEC, *Fiduciary Interpretation*, n. 27.
[19] SEC, *Fiduciary Interpretation*, 9.
[20] SEC, *Fiduciary Interpretation*, 9.
[21] SEC, *Fiduciary Interpretation*, 8.

advisers be designed to give tailored advice to each individual client, rather than programmatic advice designed for clients with similar investment objectives but who may have very different individual circumstances. Algorithms that generate goal-based investment advice without regard for each client's individual circumstances might not be consistent with the Interpretation.

The Fiduciary Interpretation does not provide specific guidance on how robo-advisers should fulfill their fiduciary duties. However, the SEC issued staff guidance on robo-advisers in 2017 and the Fiduciary Interpretation refers to that guidance.[22]

7.2.4 SEC Staff Guidance

The SEC's Division of Investment Management (2017) issued regulatory guidance on robo-advisers (the Guidance).[23] The Guidance states that the staff has been monitoring and engaging with robo-advisers to evaluate how they meet their obligations under the Advisers Act.

In keeping with the SEC's principles-based approach under the Advisers Act, the Guidance does not impose any obligations or requirements on robo-advisers but rather makes recommendations that robo-advisers "may wish to consider" in complying with the act.

The Guidance states that robo-advisers "should keep in mind certain unique considerations as they seek to meet their legal obligations under the Advisers Act." In recognition that not all robo-advisers are alike, the Guidance states there may be a variety of means for a robo-adviser to meet its obligations to its clients under the act.

The Guidance focuses on three areas as raising unique issues for robo-advisers: disclosures, the obligation to obtain information from clients to support the robo-adviser's duty to provide suitable advice, and compliance programs.

7.2.4.1 Disclosures

With respect to disclosures, the Guidance says that robo-advisers may wish to consider the most effective way to communicate to their clients the

[22] SEC, *Fiduciary Interpretation*, n. 27.
[23] "Staff Guidance" or "Guidance".

limitations, risks, and operational aspects of their advisory services and should consider providing the following information:

- A statement that an algorithm is used to manage individual client accounts;
- A description of the algorithmic functions used to manage client accounts (e.g., that the algorithm generates recommended portfolios; that individual client accounts are invested and rebalanced by the algorithm);
- A description of the assumptions and limitations of the algorithm used to manage client accounts (e.g., if the algorithm is based on modern portfolio theory, a description of the assumptions behind and the limitations of that theory);
- A description of the particular risks inherent in the use of an algorithm to manage client accounts (e.g., that the algorithm might rebalance client accounts without regard to market conditions or on a more frequent basis than the client might expect; that the algorithm may not address prolonged changes in market conditions);
- A description of any circumstances that might cause the robo-adviser to override the algorithm used to manage client accounts (e.g., that the robo-adviser might halt trading or take other temporary defensive measures in stressed market conditions);
- A description of any involvement by a third party in the development, management, or ownership of the algorithm used to manage client accounts, including an explanation of any conflicts of interest such an arrangement may create (e.g., if the third party offers the algorithm to the robo-adviser at a discount, but the algorithm directs clients into products from which the third party earns a fee);
- An explanation of any fees the client will be charged directly by the robo-adviser, and of any other costs that the client may bear either directly or indirectly (e.g., fees or expenses clients may pay in connection with the advisory services provided, such as custodian or mutual fund expenses; brokerage and other transaction costs);
- An explanation of the degree of human involvement in the oversight and management of individual client accounts (e.g., that investment advisory personnel oversee the algorithm but may not monitor each client's account);
- A description of how the robo adviser uses the information gathered from a client to generate a recommended portfolio and any limitations

(e.g., if a questionnaire is used, that the responses to the questionnaire may be the sole basis for the robo-adviser's advice; if the robo-adviser has access to other client information or accounts, whether, and if so, how, that information is used in generating investment advice); and
- An explanation of how and when a client should update information he or she has provided to the robo-adviser.

The Staff Guidance cautions robo-advisers to avoid misleading clients about the nature of the services they provide and avoid implying that:

- The robo-adviser is providing a comprehensive financial plan if it is not in fact doing so (e.g., if the robo-adviser does not take into consideration a client's tax situation or debt obligations, or if the investment advice is only targeted to meet a specific goal—such as paying for a large purchase or college tuition—without regard to the client's broader financial situation);
- A tax-loss harvesting service also provides comprehensive tax advice; or
- Information other than that collected by the questionnaire (e.g., information concerning other client accounts held with the robo-adviser, its affiliates or third parties; information supplementally submitted by the client) is considered when generating investment recommendations if such information is not in fact considered.

Notably, the Guidance acknowledges that robo-advisers may provide investment advice "without regard to the client's broader financial situation," thus implicitly acknowledging that robo-advisers do not necessarily give personalized investment advice tailored to each customer's complete financial circumstances.

Because robo-advisers rely on online disclosures, the staff said that unique issues may arise when they communicate key information, risks, and disclaimers. Accordingly, the Guidance reminds robo-advisers to carefully consider whether their written disclosures are designed to be effective, e.g., "not buried or incomprehensible." In particular, robo-advisers "may wish to consider":

- Whether key disclosures are presented prior to the sign-up process so that information necessary to make an informed investment decision is available to clients before they engage, and make any investment with, the robo-adviser;

- Whether key disclosures are specially emphasized (e.g., through design features such as pop-up boxes);
- Whether some disclosures should be accompanied by interactive text (e.g., through design features such as tool-tips or other means to provide additional details to clients who are seeking more information (e.g., through a "Frequently Asked Questions" section); and
- Whether the presentation and formatting of disclosure made available on a mobile platform have been appropriately adapted for that platform.

7.2.4.2 Suitability of Recommendations

The SEC Guidance requires robo-advisers, like other advisers, to provide investment advice "consistent with the fiduciary duty" they owe their clients.[24] In this regard, the Guidance reiterates that an investment adviser's fiduciary duty includes an obligation to act in the best interests of its clients and to provide only suitable investment advice. As stated in the Guidance, "an investment adviser must make a reasonable determination that the investment advice provided is suitable for the client based on the client's financial situation and investment objectives."

However, while acknowledging that robo-advisers do not necessarily perform this function, the Guidance does not state that robo-advisers will be found to have violated their fiduciary duty to provide suitable investment advice if they fail to take into consideration the client's broader financial situation.

The Guidance notes that robo-adviser questionnaires vary in the types of information they request and that some questionnaires are not designed to provide a client with the opportunity to give additional information or context concerning the client's selected responses. In addition, the Guidance notes that robo-advisers may not be designed so that advisory personnel may ask follow-up or clarifying questions about a client's responses, address

[24] See Division of Investment Management (2017), Staff Guidance, 10. ("As SEC-registered investment advisers, robo-advisers are subject to all of the requirements of the Advisers Act, including the requirement that they provide advice consistent with the fiduciary duty they owe to their clients.")

inconsistencies in client responses, or provide a client with help when filling out the questionnaire.

In a bulletin to investors who use robo-advisers, the SEC's staff indicated that robo-advisers might not meet the suitability standard under the Advisers Act. The staff cautioned that robo-advisers do not necessarily collect sufficient information to make suitable investment recommendations:

> [A] robo-adviser's recommendation is limited by the information it requests and receives from you, typically through an online questionnaire. It is important to keep in mind that some robo-advisers may obtain and consider only limited information about you (Securities and Exchange Commission 2017).
> An automated investment tool [i.e., robo-adviser] may not assess all of your particular circumstances, such as your age, financial situation and needs, investment experience, other holdings, tax situation, willingness to risk losing your investment money for potentially higher investment returns, time horizon for investing, need for cash, and investment goals. Consequently, some tools may suggest investments (including asset-allocation models) that may not be right for you.
> For example, an automated investment tool may estimate a time horizon for your investments based only on your age, but not take into account that you need some of your investment money back in a few years to buy a new home. In addition, automated tools typically do not take into account that your financial goals may change (Securities and Exchange Commission and Financial Industry Regulatory Authority 2015).

Given robo-advisers' limited interaction with customers, the Guidance suggests that a robo-adviser, when considering whether its questionnaire is designed to elicit sufficient information to support its suitability obligation, "may wish to consider" factors such as:

> Whether the questions elicit sufficient information to allow the robo-adviser to conclude that its initial recommendations and ongoing investment advice are suitable and appropriate for that client based on his or her financial situation and investment objectives;
> Whether the questions in the questionnaire are sufficiently clear and/or whether the questionnaire is designed to provide additional clarification or examples to clients when necessary (e.g., through the use of design features, such as tool-tips or popup boxes); and
> Whether steps have been taken to address inconsistent client responses, such as: – Incorporating into the questionnaire design features to alert a client

> when his or her responses appear internally inconsistent and suggest that the client may wish to reconsider such responses; or – Implementing systems to automatically flag apparently inconsistent information provided by a client for review or follow-up by the robo-adviser.[25]

The Guidance goes so far as to suggest that robo-advisers may result in unsuitable investment advice:

> Many robo-advisers give clients the opportunity to select portfolios other than those that they have recommended. Some robo-advisers do not, however, give a client the opportunity to consult with investment advisory personnel about how the client-selected portfolio relates to the client's stated investment objective and risk profile, and its suitability for that client. This may result in a client selecting a portfolio that the robo-adviser believes is not suitable for the investment objective and risk profile the robo-adviser has generated for the client based on his or her questionnaire responses.

To deal with this problem, the Guidance recommends that a robo-adviser, consistent with its obligation to act in its client's best interest, "should consider providing commentary as to why it believes particular portfolios may be more appropriate for a given investment objective and risk profile" and "may wish to consider whether pop-up boxes or other design features would be useful to alert a client of potential inconsistencies between the client's stated objective and the selected portfolio."

7.2.4.3 Compliance

The SEC Staff Guidance cautions robo-advisers that the provision of advisory services over the Internet with limited human interaction may create or accentuate risk exposures for the robo-adviser that should be addressed through written policies and procedures. The Guidance recommends that robo-advisers consider whether to adopt and implement written policies and procedures addressing areas such as:

> The development, testing, and back-testing of the algorithmic code and the post-implementation monitoring of its performance (e.g., to ensure that the code is adequately tested before, and periodically after, it is integrated

[25] DIM, *Staff Guidance*, 6–7.

into the robo-advisers' platform; the code performs as represented; and any modifications to the code would not adversely affect client accounts);
The questionnaire eliciting sufficient information to allow the robo-adviser to conclude that its initial recommendations and ongoing investment advice are suitable and appropriate for that client based on his or her financial situation and investment objectives;
The disclosure to clients of changes to the algorithmic code that may materially affect their portfolios;
The appropriate oversight of any third party that develops, owns, or manages the algorithmic code or software modules utilized by the robo-adviser;
The prevention and detection of, and response to, cyber security threats;
The use of social and other forms of electronic media in connection with the marketing of advisory services (e.g., websites; Twitter; compensation of bloggers to publicize services; "refer-a-friend" programs); and
The protection of client accounts and key advisory systems.[26]

7.2.5 *SEC Investor Alert*

Concurrent with the issuance of the staff Guidance on robo-advisers, the SEC's Office of Investor Education and Advocacy issued an Investor Alert designed to "provide individual investors with information they may need to make informed decisions if they consider using robo-advisers" (Securities and Exchange Commission 2017). The Investor Alert, combined with the Staff Guidance, indicates that the SEC is largely leaving investors on their own to act in their own best interests when receiving investment advice from a robo-adviser. The Investor Alert cautions investors that:

> While robo-advisers have similarities to traditional investment advisory programs, there are also differences. Before making a decision about whether to invest through a robo-adviser, or in deciding which robo-adviser might be best for you, you should do your own research. Make sure the robo-adviser and the investment portfolio it puts together for you are a good match for your investment needs and goals, and that you understand the potential costs, risks, and benefits of using that particular robo-adviser.

The Investor Alert highlights certain issues that investors should consider, including:

[26] DIM, 8.

- The level of human interaction important to the investor;
- The information the robo-adviser uses in formulating recommendations;
- The robo-adviser's approach to investing;
- The fees and charges involved.

With respect to the information used by robo-advisers in formulating recommendations, the Investor Alert cautions investors that a robo-adviser's recommendation is limited by the information it requests and receives and that some robo-advisers may obtain and consider only limited information about the customer. The Alert suggests that customers ask:

- Would you use the robo-adviser for a specific financial goal (for example, retirement, buying a home, or investing for your children's education), or to meet your overall financial needs more broadly?
- Does the robo-adviser's recommendation take into account your purpose in using the robo-adviser?
- Does the robo-adviser's recommendation take into account relevant personal financial information, given your goal? For example, does the robo-adviser ask for information about high interest credit card debt or student loans you may have? Does it take into account your bank and savings accounts? Does it take into account your real estate holdings, such as your home, or other investments such as retirement accounts? Does it take into account other assets that you have?
- How does the robo-adviser take into account your tolerance for risk? How you respond to the robo-adviser's questions about risk can affect what portfolio the robo-adviser recommends. In addition to the initial makeup of your portfolio, how does your risk tolerance impact how the robo-adviser might rebalance your portfolio (for example, in the event of a market decline)?

The Investor Alert notes that different robo-advisers have different approaches to investing and advises investors to take the time to understand how the robo-adviser develops a portfolio recommendation and to ask the following:

- Does the robo-adviser offer a limited range of investment products, such as only ETFs? Are the investment products utilized by the robo-adviser appropriate for your goals?
- Does the robo-adviser only offer certain limited portfolios within those investment products? How many different portfolios could your money possibly be invested in? What portfolio does the robo-adviser recommend for you and why?
- What type of accounts does the robo-adviser manage? For example, does the robo-adviser manage individual retirement accounts (IRAs)? Taxable accounts? 401(k) accounts or college savings plans?
- How does the robo-adviser handle volatility? For example, does the robo-adviser have the ability to freeze sales (i.e., not let you sell your investments for cash for a certain period of time)?
- How often is your account rebalanced? Rebalancing can have tax implications, depending on the type of account. What would trigger a change in the asset allocation or investment categories of your portfolio?
- Does the robo-adviser utilize tax-loss harvesting?

With respect to fees, the Investor Alert notes that a robo-adviser "may offer lower-cost investment advice, but if the robo-adviser utilizes investment products with high costs, your total overall costs could still be high." The Alert advises investors to ask:

- What fees would you be charged directly by the robo-adviser? Are there any other costs (e.g., brokerage fees, management fees for ETFs purchased for your account) that you would pay directly or indirectly?
- How is the robo-adviser compensated? Does the way it is compensated create any conflicts of interest with you, the investor? For example, is the robo-adviser paid to offer particular products or does it offer only products with which it is affiliated (e.g., mutual funds sponsored by the robo-adviser or its affiliates)?
- Are there penalties or fees if you want to withdraw your investment, or transfer or close your account? Liquidating an account may have tax implications for you as well.
- Does the amount you are charged depend on how much money you invest?
- Can the costs and fees change over time?

- Does the robo-adviser pay a referral or marketing fee, or other incentives for finding new clients? Robo-advisers may use different marketing techniques, such as paying money to others or providing discounted fees for making client referrals. You should understand if a robo-adviser has that kind of feature, even if you are not paying a fee yourself.

It is unclear to what extent the SEC believes that investors will heed this cautionary advice when using with robo-advisers. Robo-advisers are not required to give their customers a copy of the Investor Alert and it is unclear how the SEC anticipates making investors aware of it other than by posting it on its website. Retail investors do not generally visit the SEC's website.

7.2.6 *Supervision*

Investment advisers are not subject to regular or frequent inspection by SEC examiners. An SEC staff study reported that the average registered adviser could expect to be examined less than once every 11 years.[27] The SEC has not indicated the extent to which it has conducted supervisory or compliance inspections of robo-advisers.

No self-regulatory authority exists for investment advisers, unlike broker-dealers who are required to be members of and are regulated by the Financial Industry Regulatory Authority (FINRA).

7.3 Securities Exchange Act of 1934

The Securities Exchange Act of 1934 (Exchange Act) regulates securities broker-dealers. Robo-advisers typically offer their advisory services in conjunction with a broker-dealer, which executes securities transactions for a robo-adviser's customer accounts. The broker-dealer may or may not be an affiliate of the robo-adviser. However, robo-advisers themselves generally are not broker-dealers and are not required to register as such.

A "broker" is defined in the Exchange Act to mean "any person engaged in the business of effecting transactions in securities for the account of

[27] See Securities and Exchange Commission (2011), 14. In contrast, the study found that the Financial Industry Regulatory Authority (FINRA) examined 54 percent of its broker-dealer members in 2009.

others."[28] The SEC has taken the position that an investment adviser is not engaged in "effecting" securities transactions and is not required to register as a broker-dealer merely because it has discretionary authority to place orders with brokers and to execute securities transactions for client accounts without specific compensation for this function.[29]

7.4 Investment Company Act of 1940

The Investment Company Act of 1940 regulates investment companies, commonly known as mutual funds. The act also has been interpreted broadly by the SEC to include advisory programs that provide similar investment advice to groups of individuals, raising an issue as to whether robo-advisers are investment companies.

The Investment Company Act defines "investment company" generally as an issuer of securities that "holds itself out as being engaged primarily ... in the business of investing, reinvesting, or trading in securities."[30] The SEC interprets this definition to include certain advisory programs that invest client assets on a group basis. In 1997, the SEC adopted a rule-Rule 3a-4—creating a safe-harbor exclusion for such programs that meet certain requirements.

Rule 3a-4 provides a nonexclusive safe harbor from the definition of "investment company" for certain investment advisory programs designed to provide the same or similar professional portfolio management services on a discretionary basis to a large number of advisory clients having relatively small amounts to invest.[31] An investment advisory program that is organized and operated in accordance with the rule's provisions is not required to register as an investment company under the Investment Company Act or to comply with the act's requirements.[32] Robo-advisers may rely on this rule in seeking to avoid investment company regulation.

[28] Investment Advisers Act of 1940, 15 U.S.C. § 78c(a)(4)(A)(2020).

[29] 50 Fed. Reg. 1 49835, 49839 (1985).

[30] Investment Company Act of 1940, Pub. L. No. 111-72 § 3(a)(1) (2009).

[31] Investment Company Act of 1940, 17 C.F.R. § 270 (3a)(4) (1997), Status of Investment Advisory Programs.

[32] The rule is intended to be a nonexclusive safe harbor; a program that is not organized and operated in a manner consistent with the rule does not necessarily meet the Investment Company Act's definition of "investment company."

In order for the safe harbor to apply, Rule 3a-4 provides that:

(i) each client's account must be managed on the basis of the client's financial situation and investment objectives, and in accordance with any reasonable restrictions imposed by the client on the management of the account;
(ii) the sponsor of the program must obtain sufficient information from each client to be able to provide individualized investment advice to the client;
(iii) the sponsor and portfolio manager must be reasonably available to consult with each client;
(iv) each client must have the ability to impose reasonable restrictions on the management of the client's account;
(v) each client must be provided with a quarterly account statement containing a description of all activity in the client's account; and
(vi) each client must retain certain indicia of ownership of all securities and funds in the account.

In proposing the rule, the SEC stated that it was intended to cover investment advisory programs where a client's account is managed on a discretionary basis in accordance with "pre-selected investment objectives' and where "clients with similar investment objectives often receive the same investment advice and may hold the same or substantially the same securities in their accounts." The SEC stated that, in light of the similarity of management, "some of these investment advisory programs may meet the definition of investment company under the Investment Company Act."[33]

The SEC noted that clients of an investment advisory program with similar investment objectives may hold substantially the same securities in their accounts in accordance with a portfolio manager's model, but that this "does not necessarily indicate that clients in the program have not received individualized treatment for purposes of the rule." The SEC also stated that it would not be necessary under the rule for a portfolio manager to make separate determinations regarding the appropriateness of each transaction for each client prior to effecting the transaction.

[33] Company Act, 17 C.F.R. § 270 (3a)(4); Investment Company Act of 1940, 17 C.F.R. § 270 (3a)(4) (1995).

The SEC also noted that, prior to the rule's adoption, the Division of Investment Management had issued no-action letters allowing programs that allocate client assets in accordance with computerized investment allocation models to operate without being deemed investment companies.[34]

With respect to the requirement for initial and ongoing client contact, the SEC stated that a program relying on the rule must provide that the sponsor or a designated person contact and solicit information when the client opens the account and at least annually to determine whether there have been changes in the client's financial situation or investment objectives. The SEC stated that the requirement for client contact "is critical to the provision of individually tailored advice." However, the SEC stated that the rule does not dictate the manner in which a sponsor contacts its clients annually and "contact can be made, for example, in person, by telephone, or by letter or electronic mail that includes a questionnaire requesting the client to provide or update relevant information."

Depending on how a robo-adviser is structured, it may or may not qualify for the safe harbor from investment company regulation. It is unclear, for example, whether the safe harbor would apply to programs that do not permit the customer to purchase and own individual securities. Moreover, to the extent robo-advisers do not manage client accounts on the basis of each client's financial situation and clients do not have reasonable access to personnel who are available to consult with the client, the safe harbor might not be available. Robo-advisers are designed to provide investment advice based on asset allocation formulas and strategies that result in the same investment recommendations to investors with broadly similar investment goals. Thus, the advice may not be based on each client's individual financial situation. Robo-advisers are designed to operate with no individual account manager and limit communication with their clients to an Internet interface. Thus, clients might not be able to consult with the sponsor or personnel of the manager of the client's account who are knowledgeable about the account.

[34] The SEC cited no action letters to Qualivest Capital Management Inc. (pub. avail. July 30, 1990) (sponsor will use computerized investment allocation model to allocate and reallocate client assets among money managers); Atlantic Bank of New York (pub. avail. June 7, 1991) (sponsor's asset allocation recommendation will be based on client's investment needs and sponsor's model portfolios).

The SEC's staff, in issuing its Guidance on robo-advisers in 2017, stated that robo-advisers should consider whether they meet the conditions of Rule 3a-4 and contact the staff for further guidance if necessary:

> Robo-advisers should consider whether the organization and operation of their programs raise any issues under the other federal securities laws, including the Investment Company Act of 1940, and in particular Rule 3a-4 under that act. To the extent that a robo-adviser believes that its organization and operation raise unique facts or circumstances not addressed by Rule 3a-4, such adviser may wish to consider contacting the staff for further guidance.[35]

As of this date, the SEC does not appear to have found that any robo-adviser is operating as an unregistered investment company.

7.5 Employee Retirement Income Security Act of 1974

In addition to the statutes discussed above, robo-advisers need to be mindful of the Employee Retirement Income Security Act of 1974 ("ERISA"), which imposes fiduciary obligations on "fiduciaries" of employee benefit plans, including 401(k) plans and individual retirement accounts (IRAs) pursuant to the Internal Revenue Code.

Robo-advisers that provide investment advice to retirement investors generally are "fiduciaries" under ERISA. A person or firm is a "fiduciary" to the extent it renders investment advice with respect to assets of a plan or IRA and receives a fee or other compensation, direct or indirect, for doing so.[36]

ERISA imposes a number of duties on fiduciaries, including a duty of loyalty, a duty to act for the exclusive purposes of providing plan benefits and defraying reasonable expenses, and a duty of care.[37] In addition, the act prohibits plan fiduciaries from engaging in certain "prohibited"

[35] See Division of Investment Management (2017), *Staff Guidance*, 2.

[36] Employee Retirement Income Security Act of 1974, Pub. L. No. 116–136 § 3(21)(A)(ii) (2020); Internal Revenue Code of 1986, 26 U.S.C. § 4975(e)(3) (2020), as amended provides a similar definition of "fiduciary" for purposes of Code section 4975 dealing with individual retirement accounts (IRAs).

[37] ERISA, Pub. L. No. 116–136 § 404(a). The DOL has said the duty of care is grounded in the prudent man standard from trust law.

transactions.[38] Fiduciaries are personally liable for losses sustained by a plan which result from a violation of these rules.[39]

In 2006, Congress created an exemption from the prohibited transaction provisions of ERISA for certain "eligible investment advice arrangements" meeting fee-leveling requirements or using computer models. Robo-advisers are eligible for this exemption, although not all robo-advisers can meet its conditions. The exemption applies to advice arrangements that utilize a computer model designed and operated to:

(A) Apply generally accepted investment theories that take into account the historic risks and returns of different asset classes over defined periods of time, although nothing herein shall preclude a computer model from applying generally accepted investment theories that take into account additional considerations;
(B) Take into account investment management and other fees and expenses attendant to the recommended investments;
(C) Appropriately weigh the factors used in estimating future returns of investment options;
(D) Request from a participant or beneficiary and, to the extent furnished, utilize information relating to age, time horizons (e.g., life expectancy, retirement age), risk tolerance, current investments in designated investment options, other assets or sources of income, and investment preferences; provided, however, that nothing herein shall preclude a computer model from requesting and taking into account additional information that a plan or a participant or beneficiary may provide;
(E) Utilize appropriate objective criteria to provide asset-allocation portfolios comprised of investment options available under the plan;
(F) Avoid investment recommendations that:
    (1) Inappropriately favor investment options offered by the fiduciary adviser or a person with a material affiliation or material contractual relationship with the fiduciary adviser over other investment options, if any, available under the plan; or

[38] ERISA, Pub. L. No. 116–136 § 406.
[39] ERISA, Pub. L. No. 116–136 § 409.

(2) Inappropriately favor investment options that may generate greater income for the fiduciary adviser or a person with a material affiliation or material contractual relationship with the fiduciary adviser; and

(G) (1) take into account all designated investment options ... available under the plan without giving inappropriate weight to any investment option[40]

Prior to using a computer model, the fiduciary adviser must obtain a written certification from an eligible investment expert that the computer model meets the requirements of the exemption. Before providing investment advice, the fiduciary adviser also must disclose to plan participants:

(A) The role of any party that has a material affiliation or material contractual relationship with the fiduciary adviser in the development of the investment advice program and in the selection of investment options available under the plan;

(B) The past performance and historical rates of return of the designated investment options available under the plan, to the extent that such information is not otherwise provided;

(C) All fees or other compensation that the fiduciary adviser or any affiliate thereof is to receive (including compensation provided by any third party) in connection with—

  (1) The provision of the advice;

  (2) The sale, acquisition, or holding of any security or other property pursuant to such advice; or

  (3) Any rollover or other distribution of plan assets or the investment of distributed assets in any security or other property pursuant to such advice;

(D) Any material affiliation or material contractual relationship of the fiduciary adviser or affiliates thereof in the security or other property;

(E) The manner, and under what circumstances, any participant or beneficiary information provided under the arrangement will be used or disclosed;

[40] 29 C.F.R. § 2550.408g-1 (2020).

(F) The types of services provided by the fiduciary adviser in connection with the provision of investment advice by the fiduciary adviser;
(G) The adviser is acting as a fiduciary of the plan in connection with the provision of the advice; and
(H) That a recipient of the advice may separately arrange for the provision of advice by another adviser that could have no material affiliation with and receive no fees or other compensation in connection with the security or other property.[41]

The fiduciary adviser also must engage an independent auditor to audit the advisory arrangement's compliance with the Department of Labor (DOL) rule.

It is questionable whether this exemption is available to robo-advisers that provide only goal-based investment advice. As noted, the exemption requires that the computer model request information from the user on current investments and other assets or sources of income and contemplates that the computer model will generate comprehensive advice, something robo-advisers typically do not do. Moreover, some robo-advisers use computer models that favor investment options offered by the robo-adviser or its affiliates and thus would not be eligible for the exemption.

Some robo-advisers may rely on another exemption from ERISA's prohibited transaction rules for "fee leveling" arrangements. Such an arrangement is one in which any investment advice is based on "generally accepted investment theories that take into account the historic risks and returns of different asset classes over defined periods of time" as well as other considerations. The advice also must take into account investment management and other fees and expenses attendant to the recommended investments and also "to the extent furnished" by a plan, participant, or beneficiary:

> information relating to age, time horizons (e.g., life expectancy, retirement age), risk tolerance, current investments in designated investment options, other assets or sources of income, and investment preferences of the participant or beneficiary.[42]

[41] 29 C.F.R. § 2550.408g-1 (2020).
[42] 29 C.F.R. §2550.408g-1(b)(3).

An adviser relying on the exemption is required to request such information but the DOL's implementing rule specifically provides that nothing requires that the investment advice take into account information requested but not furnished. The adviser may not receive from any affiliate or other party, directly or indirectly, any fee or other compensation (including commissions, salary, bonuses, awards, promotions, or other things of value) that varies based on the selection of a particular investment option – the fee "leveling" provision.

The DOL in 2016 adopted a rule interpreting the definition of the term "fiduciary" as it applies to ERISA plans and individual retirement accounts (IRAs).[43] Under DOL's so-called fiduciary rule, a person is generally deemed to be a "fiduciary" if he or she provides an investment recommendation to a plan or IRA for a fee or other compensation, direct or indirect. However, the DOL's rule was vacated by an appellate court in 2018 which ruled that the DOL lacked authority to adopt the rule.[44]

7.6 State Regulation

Robo-advisers also may be subject to regulation under state securities or other laws. Massachusetts, through its Securities Division, has been the most proactive among the fifty states in addressing robo-advisers. In 2016, the Massachusetts Securities Division issued a policy statement concluding that "fully automated robo-advisers, as currently structured, may be inherently unable to carry out the fiduciary obligations of a state-registered investment adviser."[45]

7.7 Conclusion

As this chapter has described, robo-advisers are regulated in the United States under the existing regime that governs financial firms that provide investment advice to investors. In particular, robo-advisers are treated

[43] 29 C.F.R. §2510.3–21 (2020); 81 Fed. Reg. 20946 (2016) (final rule).

[44] *Chamber of Commerce of the United States v. United States Dep't of Labor*, 885 F.3d 360 (5th Cir. 2018).

[45] See Massachusetts Securities Division (2016a). *See also* Massachusetts guidance to state-registered investment advisers on how to best comply with the fiduciary duties owed to their clients when they establish sub-advisory relationships with third-party robo-advisers. Massachusetts Securities Division (2016b).

as "investment advisers" for purposes of the Investment Advisers Act of 1940 and as such generally are required to register with the Securities and Exchange Commission. Robo-advisers also are subject to regulation by the various states, although most states do not regulate them differently from other investment advisers.

The Securities and Exchange Commission has identified a number of unique regulatory issues raised by robo-advisers and has issued guidance addressing how existing regulations apply to their activities, including with respect to disclosures, suitability of recommendations, and compliance. The Commission has also issued an investor alert highlighting issues that investors should consider before using robo-advisers, including the level of human interaction important to the investor, information the robo-adviser uses in formulating recommendations, and the robo-adviser's approach to investing.

As reflected in this chapter, US regulators generally have taken the position that, while robo-advisers may present unique issues, the existing framework is adequate to regulate their activities and protect investors. No new regulatory regime has been proposed to address robo-advisers in the United States nor is any new regulatory framework anticipated.

References

Division of Investment Management. 2017. *IM Guidance Update–February 2017 No. 2017-02 – Robo-Advisers*. Tech. rep. P.8. US Securities and Exchange Commission.

Massachusetts Securities Division. 2016. *Policy Statement: Robo-Advisers and State Investment Adviser Registration*. Last Accessed 1 April 2016. http://www.sec.state.ma.us/sct/sctpdf/Policy-%20Statement%E2%80%93Robo-Advisers-and-State-Investment-Adviser-Registration.pdf.

Massachusetts Securities Division. 2016. *Policy Statement: State-Registered Investment Advisers' Use of Third-Party Robo-Advisers*. Last Accessed 1 July 2016.

Securities and Exchange Commission. 2011. *Study on Investment Advisers and Broker-Dealers*. Tech. rep. https://www.sec.gov/news/studies/2011/913studyfinal.pdf.

Securities and Exchange Commission. 2017. *Investor Bulletin: Robo-Advisers*. Investor Alerts and Bulletins. Last Accessed 20 March 2020. https://www.sec.gov/oiea/investor-alerts-bulletins/ib_roboadvisers.html.

Securities and Exchange Commission. 2019. *Commission Interpretation Regarding Standard of Conduct for Investment Advisers–Release No. IA-5248; File No. S7-07-18.* Tech. rep.

Securities and Exchange Commission and Financial Industry Regulatory Authority. 2015. *Investor Alert: Automated Investment Tools.* Investor Alerts and Bulletins. Last Accessed 20 March 2020. https://www.sec.gov/oiea/investor-alerts-bulletins/autolistingtoolshtm.html.

CHAPTER 8

Regulation of Robo-Advisory in Europe and Germany

Christian Hammer

Contents

C. Hammer (✉)
Netfonds Financial Service GmbH, Hamburg, Germany
e-mail: chammer@netfonds.de

© The Author(s) 2021

P. Scholz (ed.), *Robo-Advisory*, Palgrave Studies in Financial Services Technology, https://doi.org/10.1007/978-3-030-40818-3_8

8.1 What Are the Activities for Which a Provider Needs a License?

Hardly any other market is as regulated as the financial services market. Only a few years ago, a broad spectrum of completely or at least largely unsupervised products and services were available, but as a result of the financial crisis and various investment scandals, legislators had no choice but to tighten the regulations for products and services considerably. Today, there are hardly any investment products available that would not require a license or approval.

This has also had a profound impact on the behavior of fintech start-ups. While only a few years ago they were still trying to position and develop their companies outside government supervision, various successful start-ups have shown that strong growth and regulatory compliance are not mutually exclusive. Compliance with legal requirements is even mandatory for businesses with exit plans, as hardly any buyers would pay an ambitious price for a business model operating outside legal requirements. The resulting civil, administrative and even criminal risks would either cause buyers to back off or lead to a massive reduction in the purchase price. Both are in stark contrast to what businesses with exit plans want. In view of the increased willingness of investors to engage in disputes, compliance with the rules and regulations is also of enormous importance to all providers of digital products with capital market ambitions. Many so-called investor law firms are now scouring the market for failed offers and seeking to represent a large number of investors in schematically similar cases or potential class-action lawsuits funded by legal expenses insurance or litigation financing companies. Irrespective of the personal motivation of the start-ups, it is clear that compliance with legal requirements is crucial to a successful business model.

Depending on the business model adopted, fintech firms may also fall under the supervision of the German Federal Financial Supervisory

Authority (BaFin) in accordance with the provisions of the KWG.[1] Other providers, the so-called investment intermediaries, however, are only subject to the relevant provisions of the Trade Regulation Act. In principle, we have a regulatory "level playing field" in Germany, that is, the same financial services and the same risks are subject to the same supervision and regulation. How does this work in practice? Typically, robo-advisors operate as so-called intermediaries. Some provide their clients with advisory and management services and some operate solely as intermediaries. Robo-advisor is a generic term used to describe various technical systems used by investors to invest or manage capital themselves.[2] When in doubt, companies need to check with the relevant authorities and/or professionals to see what licenses or permits they need.

Whether a license will be required or not will depend on the investment products offered and the activities the company carries out on behalf of its clients.

8.1.1 Product-Related

Since 2017, all financial investment products are considered to be financial instruments within the meaning of the German Banking Act (KWG). Only physical direct investments, such as the purchase of gold or silver bars, do not fall under the term "financial instrument". The same applies to cooperative shares. The term financial instruments thus covers nearly all the products available to private clients on the capital market. Under the central provision of Article 32 KWG, anyone wishing to provide advisory or management services linked to financial instruments or to act as an intermediary for financial instruments needs a license from BaFin. The first sentence of Article 2 (6) No. 8 KWG creates a special exception from this strict principle for companies, which solely provide investment advice and brokerage services and do not engage in any other financial services or banking transactions (see below for more details). Although these companies meet the material requirements to be classified as financial services institutions in accordance with Article 1 (1a) KWG, they are not to be regarded as institutions within the meaning of the KWG and are,

[1] Details of various products at Bundesanstalt für Finanzdienstleistungsaufsicht (2020).

[2] For the definition of typical business models, see also Oppenheim and Lange-Hausstein (2016).

therefore, not subject to a license requirement under Article 32 KWG. The activities of these companies must be limited to the following financial instruments:

- Shares in investment funds issued by a domestic investment management company,
- EU investment assets and foreign AIFs that may be sold under the German Capital Investment Act (KAGB),
- Investments within the meaning of Article 1 (2) of the German Investment Products Act (VermAnlG).

If a company adheres to this, it will not require a license from BaFin in accordance with Article 32 (1) KWG. A business license as a financial intermediary under Article 34f (1) of the German Trade Regulation Act (GewO) is then sufficient, an exception that is unique in Europe.

Cross-border provision of services or the establishment of a branch in another EU member state or in a third country are not covered by this license under the Trade Regulation Act; this can only be done by companies licensed under Article 32 KWG. In that regard, investment intermediaries are regionally restricted to the territory of the Federal Republic of Germany.

What is hidden behind the three classes of financial instruments mentioned above? Generally speaking, one could say that investment intermediaries are "fund brokers", where the term funds refers to so-called UCITS funds and alternative investment funds such as open-end real estate funds, but also closed-end products under the KAGB such as ship funds or closed-end real estate funds. This means that most ETFs licensed under Article 34f GewO are also a permissible subject of services. However, a more detailed analysis is needed to provide a concrete response for individual ETFs.

The umbrella term ETFs also covers ETCs (exchange traded commodities) and ETNs (exchange traded notes). The legal structure is decisive for the delimitation. Investment intermediaries under Article 34f GewO may only provide advice or brokerage services for "real" ETFs within the meaning of the KAGB, which are issued as UCIT funds. Whether or not they are swap-based does not affect this. In contrast, ETCs and ETNs are merely securitized claims, i.e. bearer bonds. This is why a KWG license is required to broker these products; a 34f license will not be sufficient. To check whether this is a real ETF, we recommend reviewing the sales prospectus. Under KAGB, products requiring a prospectus are referred to as "real" ETFs and can be brokered by investment intermediaries. A license is needed for products requiring a prospectus in accordance with the German Securities Prospectus Act (WpPG). Other financial instruments permitted under the license under Article 34f GewO are so-called financial assets as defined under Article 1 (2) German Investment Act (VermAnlG). These include investments with low or no fungibility such as not-securitized shares that grant participation in the results of a company, for example, trust assets, participatory loans, subordinated loans, profit participation certificates, silent participations, registered bonds and other investments that grant or hold the prospect of interest and repayment or cash settlement assets in exchange for the temporary provision of money. These are generally less interesting for fintech firms as they can be very complex, costly, illiquid and not easily tradeable.

In view of the high level of complexity and the resulting need to assist private investors with information, only a few fintech firms have dared to approach this product group to date. Digital asset managers are generally not interested in such non-fungible forms of investment, but investment advisors appear to be caught in the point of tension between the need to clarify, on the one hand, and a swift conclusion on the other. Only in the broking business are there individual providers offering cost-effective subscription options for investments.

8.1.2 Service-Related

In addition to the products, the execution of buy/sell orders and the discretionary latitude applied is also intriguing. The investment-related activities of robo-advisors typically fall under investment advice, investment and contract broking and/or portfolio management. These activities have to be kept separate from the others, as they require different licenses and compliance with different requirements.

While the license under Article 34f GewO is sufficient to offer brokerage services and investment advice, it is necessary to obtain the license under Article 32 KWG from BaFin to provide contract broking and portfolio-management services.

8.1.2.1 Investment Advice

Under the law, investment advice is defined as providing clients or their representatives with personal recommendations in respect of transactions relating to certain financial instruments where the recommendation is based on an evaluation of the investor's personal circumstances or is presented as being suitable for the investor and is not provided exclusively via information distribution channels or for the general public.[3]

Behind this lengthy legal definition there are highly relevant questions for robo-providers. The focus here is primarily on the question of when the product is directed at clients, what qualifies as a recommendation and when the necessary check of personal circumstances or suitability is considered to be suitable for the investor.

Clients are not only persons who actually enter into a business relationship with the provider but also those who are merely interested in the product.[4] For protection purposes, categorization as a client therefore begins at an early stage. It is sufficient to use any targeted approach aimed at concluding a contract at a later date, that is, any advertising measure.[5] Every visitor to a homepage is therefore considered to be a prospect and therefore a "client".

A "recommendation" is when the client is advised that a particular action is in his interest.[6] A recommendation is a recommendation made to a client to buy, sell or not to buy a financial instrument.[7] It does not matter whether the client subsequently acts upon that recommendation or not.

A clear distinction must be made between recommendation and information where the company merely provides its clients with information about their assets invested in financial instruments without making any

[3] Gesetz über das Kreditwesen, BGBl. I page 881

[4] This is clear from the legal definition under which clients are any natural or legal persons for whom investment firms provide or arrange investment services or ancillary investment services. The MiFID Directive 2004/39/EC already extended the information and exploration obligations provided for under Article 19 (2) to "prospective clients".

[5] This has already been set out in the predecessor provision with identical wording, Gesetz über den Wertpapierhandel, BGBl. I § 543.31(a)(2020), Schwark and Zimmer (2010), Wertpapierhandel, § WpHG, paragraph 5

[6] Cf. Joint information sheet issued by BaFin and the Deutsche Bundesbank on investment advice, version 2/2019.

[7] European Parliament and of the Council, 2006/73/EC § 52 (2006) implementing Directive 2004/39/EC.

concrete proposals to change the composition of those assets.[8] Practical difficulties usually arise for robo-providers in the context of their online offerings, for example, when certain financial instruments are presented and their advantages or opportunities explained or when certain financial instruments are advertised. The client's point of view is always decisive here. If this person has the impression that this goes beyond the mere provision of information and promotion of the product by suggesting to buy or sell the product, this will typically be construed as a recommendation within the meaning of the legal definition.

This recommendation must relate to "transactions in specific financial instruments", that is, not just to a particular asset class.[9] According to BaFin, the typical robo-advice fulfills this criterion if the client receives investment proposals relating to certain financial instruments, such as certain funds or securities and not just asset classes.[10] Where a product only generates an asset allocation (e.g. European equities 30%, US equities 20%, emerging markets equities 15%, bonds 20%, cash 15%), this does not qualify as investment advice, as no recommendations are made on specific financial instruments. However, if the client also receives concrete recommendations on financial instruments and asset classes, then this satisfies the investment advice requirements. This also applies if different product alternatives are recommended, and it is up to the client to make the selection. In our view, the vast majority of robo-advisors work with concrete proposals for specific financial instruments.

As a rule, the decisive factor tilting the scales for or against investment advice is whether the recommendation given reflects the "personal circumstances of the investor" or it is presented as "suitable for the investor". Few robo-providers would operate today without querying the personal criteria of the prospective client online and creating an investment proposal tailored to the client's communicated needs. In most cases, providers not only ask for the investment amount and the investment horizon but above all the risk tolerance of the prospective clients. Information on the financial

[8] Cf. Joint information sheet issued by BaFin and the Deutsche Bundesbank on investment advice, version 2/2019.

[9] Entwurf eines Gesetzes zur Umsetzung der Richtlinie über Märkte für Finanzinstrumente und der Durchführungsrichtlinie der Kommission (Finanzmarkt-Richtlinie-Umsetzungsgesetz), BT-Drs. 16/4028 § 12 (2007).

[10] Cf. Joint information sheet issued by BaFin and the Deutsche Bundesbank on investment advice, version 2/2019.

circumstances of the prospective clients is also collected. This means that these providers undoubtedly fulfill the criterion of "taking into account personal circumstances" in their recommendations, even if the investment proposal is generated by an algorithm.

The joint information sheet of BaFin and Deutsche Bundesbank on investment advice thus focuses on the "financial situation" of the investor. It is, therefore, sufficient for the provider to obtain only relatively general information, to the extent that it then recommends concrete financial instruments on this basis.

In addition, it is also sufficient to be perceived as having taken into account the personal circumstances of the investor when making the recommendation. This will normally be the case if the client believes that the investment proposal reflects the information provided, irrespective of whether this applies in full or in part.[11]

This can be counteracted by using clear and easily understandable disclaimers, which have to be accepted online.[12] On this basis, the client must accept that no "individual" or "tailored" investment proposals will be provided. Investment advice can be provided with a license in accordance with Article 32 KWG or as an investment intermediary. The main differences are product-related (see above) but can also be found in some details relating to their duty of care.

8.1.2.2 Investment Broking

The definition of investment broking is satisfied when companies act as intermediaries and their business involves the buying and selling of financial instruments.[13] The broker acts as an intermediary that passes on the investor's declaration of intent to buy or sell financial instruments to the person with whom the investor wishes to conclude such a transaction.[14] In this case, it is irrelevant whether the forwarded declaration of intent is an

[11] Cf. Tertilt and Scholz (2018). It has been found that robo-advisors often use comparatively little information from onboarding to make a portfolio recommendation.

[12] CESR's Technical Advice on Possible Implementing Measures of the Directive 2004/39/EC on Markets in Financial Instruments, CESR 05-024c, 9 (2005).

[13] See BaFin's Guidance Notice on the subject of investment broking at Bundesanstalt für Finanzdienstleistungsaufsicht (2017b).

[14] This administrative practice has also been confirmed by the Federal Court of Justice (BGH). Bundesgerichtshof, WM 2014, 221/222, III ZR 73/12.

offer by the investor to the seller or the acceptance of an offer by the seller. The form of transmission also does not matter.

Accordingly, the transmission can also take place electronically, for example through a computer system. In accordance with the administrative practice of the supervisory authorities, investment intermediaries are those who provide an IT system which is used to pass on an investor's declarations of intent to buy or sell financial instruments to any contracting parties. If, as part of the robo-service, the order is placed through the IT system of the provider, the latter will provide the investment-broking service.

In addition, investment brokerage also involves targeted encouragement of the investor to get them to conclude a transaction with a third party to buy or sell financial instruments. Accordingly, investment broking is also provided by those who consciously and finally influence an investor to conclude a transaction on the acquisition or sale of financial instruments.[15] Anyone who merely brokers contact between the investor and a seller of financial instruments is not involved in investment broking if he or she makes said connection in the form of a mere referral service within the meaning of the German Trade Regulation Act.[16] In practice, it is difficult to distinguish between investment broking requiring a license and referral broking not requiring a license. The decisive criterion here is whether the client's will to conclude a transaction is influenced in any way. On the part of the authorities, there is certainly the tendency to interpret the notion as applying to intermediary activities requiring a license at the expense of activities that do not require a license. This is because (irrespective of whether this is considered to be investment advice or not) the well-known robo-providers typically propose very specific transactions; it might be reasonably expected that the brokerage activity requires a license, even if the provider's IT system is not used for the specific orders.

According to the settled case-law, the broking of asset-management contracts is not considered to be investment broking.[17] Asset management is aimed at buying and selling financial instruments for the account of the investor on the basis of discretionary investment decisions. Therefore, both the forwarding of a declaration of intent aimed at the conclusion of an asset-management agreement and the influencing of an investor

[15] See Bundesanstalt für Finanzdienstleistungsaufsicht (2017b).
[16] Schaefer, in Boos et al. (2016); Gesetz über das Kreditwesen, § 1 KWG, paragraph 136.
[17] European Court of Justice, Z.I.P. 1362 (2017).

to conclude an asset-management agreement are not considered to be investment.- broking activities. Similar to investment advice, investment-broking services can also be provided both under the license pursuant to Article 32 KWG and Article 34f GewO.

8.1.2.3 Contract Broking

In contrast to investment broking, contract broking requires a license from Bafin in accordance with Article 32 KWG. Contract broking refers to buying and selling financial instruments in the name of and for the account of a third party. A contract broker also acts as an intermediary. Whereas the investment broker passes on the client's declaration of intent to the buyer or seller of the financial instruments in the capacity of an intermediary, the contract broker submits their own declaration of intent as their client's representative. These services are thus mutually exclusive. Investment broking is more common in practice.

8.1.2.4 Portfolio Management

Under Article 1 (1a) sentence 2 No. 3 KWG, portfolio management is defined as the management of individual portfolios of financial instruments for others on a discretionary basis. The distinguishing feature of portfolio management is that the individual transactions are not based on client orders. These are initiated by the providers (managers) at their discretion. The implementation of investment decisions is the responsibility of the manager rather than the client. In order to manage the securities account, the client grants the manager so-called discretionary authority. This discretionary authority gives the manager access to the client's securities account at the custodian bank.

In the meantime, some market participants, which offer their own asset-management services, are also involved in the market as fintech firms or robo-advisors. There are considerable differences between the products and services on offer. There are conventional providers who offer digital client onboarding, while at the same time they use algorithms to select and propose an investment strategy. In contrast to robo-advisors that offer investment advice, the client does not only receive a one-off (or continuous) investment recommendation. Instead, the investment portfolio is managed on an ongoing basis in accordance with investment guidelines. The transactions are also frequently based on an algorithm. Crucially, the provider does not only make investment decisions purely mathematically based on clear rules agreed in advance with the client, but also has discretionary authority (within the agreed investment guidelines).

Insofar as the investment decisions are purely mathematical and are thus strictly based on calculations, no discretion is involved. Depending on their structure, such providers will routinely provide contract-broking services. They will then also need a license under Article 32 KWG, as contract broking and investment broking are mutually exclusive.

8.1.3 Liability Umbrella as an Alternative to Holding a License

An alternative to obtaining one's own license pursuant to Article 32 KWG or Article 34f GewO is to act as so-called tied agent pursuant to Article 2 (10) KWG. Tied agents provide financial services exclusively for the account, on behalf of and under the liability of a financial institution that is in possession of a BaFin license. They use this institution's license under Article 32 KWG and are thus not subject to the product-specific restrictions of investment intermediaries. They are permitted to provide investment advice and investment-broking services, but not contract-broking and/or portfolio-management services.

The main prerequisite for acting as a tied agent is the assumption of liability by a financial institution with its own investment broking and investment advice license. Liability is assumed under a civil law agreement between the broker and the liable company. In addition, the broker has to carry out an activity in the name of and for the account of the liable company. This is only the case if the broker acts as a representative or agent of the liable company.[18] The broker concludes a transaction on behalf of the liable company as its representative, which collects the fee. The companies then agree to share any income in a contractually defined ratio. To act as tied agents, companies must notify BaFin and have to register in BaFin's register of tied agents. By operating under a liability umbrella, tied agents can focus on their own business model, while the liable financial institution is responsible for compliance with regulatory and organizational requirements and its ongoing monitoring.[19] This model is certainly appealing for start-ups. In the meantime, a number of fintech

[18] Schaefer in Boos et al. (2016), Article 2, paragraph 128.

[19] This obligation is laid down in Article 25e KWG. Gesetz über das Kreditwesen, § 25e KWG.

Table 8.1 At a glance: which asset classes could be advised and brokered

Product-related Advice, broking of	34f GewO	Portfolio manager	Bank	Liability umbrella
Equities	No	Yes	Yes	Yes
Bonds	No	Yes	Yes	Yes
Loans	No	Yes	Yes	Yes
Participation certificates	Yes	Yes	Yes	Yes
Investment funds (UCITS)	Yes	Yes	Yes	Yes
Open-end alternative investment funds	Yes	Yes	Yes	Yes
Closed-end alternative investment funds	Yes	Yes	Yes	Yes
Certificates	No	Yes	Yes	Yes

Table 8.2 At a glance: which services are related with which license

Service-related	34f GewO	Portfolio manager	Bank	Liability umbrella
Investment advice	Yes	Yes	Yes	Yes
Investment broking	Yes	Yes	Yes	Yes
Contract broking	No	Yes	Yes	Yes
Portfolio management	No	Yes	Yes	No, only advice
Broking abroad	No	Yes	Yes	Yes
Advice abroad	No	Yes	Yes	Yes
Referral broking (tipsters)	Yes	Yes	Yes	Yes

firms have opted for this solution and operate their sales platforms under the responsibility of experienced providers. At a glance see Tables 8.1 and 8.2.

8.2 Regulatory Requirements for Fintech Firms and Robo-Advisors: Practical Issues

Despite political announcements, providers of digital financial services are offered little relief or special legal regulations that deal with or even promote the specifics of their business models. Where fintech firms offer investment services, they must, like analogous providers, comply with the related duty of care obligations (developed for the latter) under the German Securities Trading Act (WpHG) or the Investment Brokerage

Regulation (FinVermV). When it comes to investment products, fintech firms must assess the appropriateness or suitability of its potential clients. The appropriateness assessment determines whether the client is in a position to evaluate a particular investment product; the suitability assessment also determines whether a specific product is suitable for this particular client and financial situation. The information and transparency requirements are extensive. Products must be tested and evaluated as part of so-called product governance.

This presents unexpected challenges to some IT-savvy start-ups counting on a quick and easy onboarding process for its clients. Few enthusiastic entrepreneurs anticipate the regulatory burden in the area of capital markets.

The decisive factor here is the type of services provided (cf. above definition). Here are some frequently problematic aspects:

8.2.1 Information Obligations

In investment and contract broking as well as in investment advice, the provider must provide the client with the necessary information to make an informed decision about the investment. This obligation exists both under supervisory and civil law. Article 63 WpHG establishes an obligation to standardize basic information. Accordingly, abstract information must be provided relating to the nature of the financial instruments and the context in which they are provided; this information should form the basis for an informed investment decision by the client. According to the wording of the WpHG, the information may also be transmitted in a standardized form. To meet the regulatory requirements, the use of brochures, for example, investment funds basics, has become established in practice, allowing providers to offer cross-product information.

However, the use of standardized information does not meet the extensive regulatory information obligations of investment firms in full, in particular, not for every financial product. By handing over the written standard information material, the financial service provider merely satisfies its regulatory obligation to provide information. In addition, providers must give the client a true and fair view (including from civil law perspective) of the opportunities and risks involved in the transaction. With respect to portfolio management, this obligation is limited to information relating to the relevant asset class,[20] whereas in the case of all other services, the obligation also extends to information on the risks of the specific investment object to be purchased. This information must be made available to the client in good time so that he can take note of it before making the investment decision;[21] subsequent transmission after the investment decision has been made is thus contrary to regulatory law and also gives rise to considerable potential liability under civil law.

It is therefore technically feasible to ensure that all relevant information is available to the client in good time to allow him to take note of it before making the investment decision. In the case of UCITS, this correlates with the obligation to provide clients with an information sheet (so-called key investor information document; KIID). In the context of investment advice to private clients, the relevant obligation applies to all types of financial instruments.

It is, therefore, mandatory for firms to provide clients with all the necessary information before they make their final investment decision. It is highly recommended to give clients the opportunity to download all the relevant information. Alternatively, the information can also be sent by email. For liability purposes, it should, in any case, be verifiable that all relevant information reached the client in good time to allow the client to become apprised of it before making the investment decision.

In addition to these risk-related information obligations, there are a number of other, mostly regulatory information obligations that need to be considered. These include conflicts of interest policy, best execution policy and the general information obligations for financial services companies.

8.2.2 Investment Broking and Execution Only Transactions

Investment and contract brokers are required to determine whether the client is in a position to understand the risks arising from the investment decision based on the client's knowledge and/or experience, based on a so-called appropriateness assessment. To this end, they ask prospective clients

[20] Schaefer, in Assmann et al. (2015), Article 23, paragraph 28; the disclosure obligations refer to share classes, markets and currencies.

[21] The question as to when a delivery is "made on time" shall be assessed on a case-by-case basis; Bundesgerichtshof, WM 2007, 1608; III ZR 145/06; the time interval between handover and subscription is crucial, as is the client's background in terms of knowledge and experience.

about their knowledge and experience with various product classes. Based on this information, providers must then check whether the investment decision is consistent with the prior knowledge and experience of investors and whether, on the basis of their knowledge, investors can accurately assess the risks associated with the investment (so called appropriateness test).

The investor is informed of the result of the assessment and can then decide whether or not to continue with the investment—even if the result of the assessment was negative. This process can be easily designed based on standardized data; an algorithm then checks the knowledge and experience. Even if a client has no experience with certain financial instruments, a transaction may still be appropriate for the client if the information was provided to the client sufficiently in advance. This certainly places higher demands on the creativity of fintech firms, as their processes are usually geared towards a fast and smooth process. Various approaches are conceivable—for example, confirmation of the client despite the lack of appropriateness to conclude transactions, but also the termination of the onboarding process and its continuation after a certain period of time.

In certain circumstances, the appropriateness assessment may even be waived altogether. In contrast to portfolio management and investment advice, regulatory law permits a reduction in the duty of care obligations for so-called execution-only transactions within the scope of investment broking. If the conditions for the execution-only transaction are met, the financial service provider's activity is limited to the mere execution of the transaction requested by the client and is not associated with any exploration, appropriateness or suitability assessment. However, the information obligations set out above must also be fulfilled in this case. The scope of the execution-only transaction is limited to cases in which all the following conditions are met: The transaction must involve "non-complex financial instruments", it must be initiated by the client who will then be informed by an explicit statement that no appropriateness assessment will be carried out.

Complex financial instruments include, for example, subordinated debt instruments and debt instruments that qualify as PRIIP (packaged retail investment and insurance-based products) within the meaning of Article 4 No. 1 Regulation (EU) No. 1286/2014 (PRIIP Regulation), as well as derivatives. Non-complex financial instruments therefore include UCITS and traditional equities and bonds traded on regulated markets. The exception for pure execution transactions can therefore be an interesting approach for pure broking platforms, allowing the providers to offer straightforward processes.

Here too, a clever and pragmatic solution will only be successful if the company has focused on the legal requirements, and the product selection derived from them. As already mentioned, the clear focus on the product side is essential. Where an undertaking is required to provide an appropriateness assessment, it must define the knowledge and experience it requires so that it might reasonably expect that the client understands the risks involved in making an investment decision. However, thanks to clearly structured onboarding processes, this does not present a major hurdle.

8.2.3 Duty of Care Obligations in Investment Advice and Portfolio Management

Providers are required to provide clients with investment advice minutes signed by the advisor (replaced now by suitability reports) after they have received investment advice and before they enter into a securities transaction. While transmission on a durable medium (e.g. by email or Postbox) is also possible, it requires the explicit consent of the client, which must be obtained in advance. In the context of automated investment advice, there is no personal advisor available to sign such minutes. Since 2018, the minutes must also explain why the recommendations made are suitable for the client. Investment advisors may only recommend to clients financial instruments and investment services that are suitable for them. This means that any recommendation must be suitable for the client. At the same time, clients must be able to understand the functionality and risks of the recommendation. For this purpose, the advisor must inquire about the client's knowledge and experience, financial circumstances and investment objectives and review his investment recommendation on the basis of the information provided. Any gaps in knowledge must be conveyed to ensure that the client understands how this works, and the opportunities and risks involved.

The assessment is summarized and explained in writing to the client in the suitability report. It describes how the recommendation was adapted to reflect the information provided by the client. For this purpose, all the information required for the suitability assessment must be incorporated in the suitability report. The expectation is that not only text modules will be used for this purpose.[22] A certain degree of individualization is therefore required. This does not quite fit in with the typical service concept of robo-advisors. In this context, however, BaFin points out that robo-advisors are not subject to any exceptions from applicable laws.[23]

The market does not seem to have found any answers to this question that would allow the recording of investment advice minutes. As ESMA has highlighted, the suitability report should be created individually and not be merely put together from readymade text modules. Some providers have therefore limited themselves to brokering asset-management services for which no such report is required, others withdrew to the field of pure broking and product information not involving personal recommendation. It is still unclear whether a solution that has been technically tailored to the respective client by the platform would be viable.

Investment advice and portfolio management are subject to far-reaching exploratory obligations. The exploration in terms of knowledge and experience, financial circumstances, investment goals and risk tolerance can be easily automated. However, compared to personal investment advice, a schematic entry form cannot determine whether further clarifying information is required from the client.

8.2.4 Practical Importance and Border Cases in the Robo-Advisory Universe

In the above text, we raise a number of practically relevant and sometimes difficult legal issues concerning robo-advisors. Next we will discuss some other practical issues providers have to deal with:

8.2.4.1 Onboarding Using Online Communication Channels

Client onboarding is subject to far-reaching legal requirements. Above all, it is crucial for providers to clearly decide which type of investment service

[22] Cf. also BaFin expert article, Suitability report: An important document for consumers, available at https://www.bafin.de/SharedDocs/Veroeffentlichungen/DE/Fachartikel/2018/fa_bj_1809_Geeignetheitserklaerung.html.

[23] BaFin, Robo-advice and auto-trading – platforms for automated investment advice and automatic trading, BaFin demands this under the so-called Minimum Compliance Requirements (MaComp). Any inconsistencies and inaccuracies in the information collected from clients should be identified and pointed out to the investor in order to resolve them 24. It is, therefore, insufficient to collect only standardized information.

they wish to offer. This is a typical conflict of interest. While prospective clients typically look for answers and solutions that meet their individual needs as far as possible, digital providers usually want IT solutions that make registration and closing quick and easy. This is the only way they can easily engage their prospective clients in a business relationship and keep the acquisition costs low. The client will be interested in receiving investment advice in the form of investment proposals tailored to his needs. On the other hand, providers often shy away from the increased expense and regulatory complexity of investment advice as well as the liability risks that may arise from unsuitable investment proposals. In particular, the obligation to hand over suitability reports (formerly advice minutes) is difficult to integrate.

Common solutions in this respect are mere recommendations of asset allocation and the submission of recommendations only after the start of a business relationship, at least in the form of the transmission of concrete data by the prospective client, but also the use of detailed disclaimers that make it clear to the client that the presented investments do not cover all relevant aspects of investment advice, and therefore, no such advice is being provided. None of the two solutions is perfect, as they both frustrate the client's desire for concrete solutions.

8.2.4.2 Hotline and Chat

This also addresses a core problem of digital providers. Many clients attach importance to human contact before they put their savings into the hands of a company. This, too, is at odds with the approach of digital providers, which tend to offer prospective clients support in the form of hotlines or chats. This in turn raises specific delineation issues and makes compliance with legal requirements more challenging. Where online communications can be clearly structured to answer the question of whether investment advice will be provided or not, this becomes much more difficult in the context of free communication. As explained, the determining criterion for or against investment advice is the consideration of the investor's personal circumstances.

An "assessment of the investor's personal circumstances" is answered in the affirmative if the client merely informs the relevant provider in general terms of his financial situation and the service provider then recommends transactions involving specific financial instruments. Alternatively, it is also sufficient that the recommendation is merely "presented as being suitable for the investor".[24] This is the case if a client can reasonably expect that the recommendation given is based on the consideration of the client's personal circumstances. The communication of personal circumstances in the course of a chat or a telephone call and the investment recommendation made online can thus overall "constitute" investment advice. The staff employed in this area of client services must therefore receive training to avoid unwanted legal consequences. This also applies to statements on products or asset classes. If a financial service provider provides its clients with incomplete or inaccurate information, the client will be entitled to claim damages and reverse the transaction. The robo-advisor, therefore, bears the investment risk. To do this, providers need to select and train suitable personnel. Nevertheless, the risks from this step into the analogue world cannot be completely eliminated.

8.2.4.3 Order Placement

The question of how a transaction is triggered again touches on the initial question of the services the company actually offers. If the order is placed outside the robo-advisor system, and the robo-advisor does not have any impact on the client's investment decision, the actual order placement is a matter between the client and the custodian. In practice, however, this will be the absolute exception, as robo-advisors usually also offer order placement. They then act as investment or contract brokers. The relevant delimitation has already been explained above.

An important question from the practical standpoint is when so-called social, copy or mirror trading models will be regarded as broking and/or management activities. The process of this form of order initiation typically runs as follows: Clients choose the trading strategy they want to follow. To copy and execute the individual trading decisions for their portfolio, they subscribe to information about changes in the respective portfolio composition. The provider then either forwards the trading decision as a client order to the client's custodian bank (Model 1) or transmits the concrete change to the client, who then orders the transaction (usually automatically through software) (Model 2).

In Model 1, the investor's assets are managed in accordance with fintech decisions. The investor grants the company the appropriate discretionary

[24] See Bundesanstalt für Finanzdienstleistungsaufsicht (2016). The custodian then executes the submitted order and records it in the securities account.

authority/powers. This also applies if fintech is backed by another provider who gives trading signals to them. The activity of this signal generator is generally attributed to the provider who places the specific orders. As a result, this platform routinely offers portfolio-management services.

Model 2 is different. Depending on the technical configuration, the client puts an order through his computer, which is then sent to the custodian for execution via the robo-advisor's IT system. By forwarding the order to the custodian, the provider typically provides the investment broking service.

If the provider "translates" the investment proposal into a client order without the client's own declaration of intent and forwards it to the custodian, the provider typically acts as the contract broker "in the name of" the client.

In this respect, the question arises as to whether the automatically generated declarations of intent are to be understood as the client's own declarations or as declarations made by the robo-advisor. Civil case law also recognizes automatically generated declarations as effective, non-actionable declarations of intent by the software owner, although submitted by "auto-reply" if the computer used only executes commands previously determined by human programming and the declaration, therefore, originates in an act initiated by the declaring party and based on its will .[25] The decisive factor is that the client can fundamentally influence whether an order is generated or not. To what extent these principles are also recognized by BaFin is uncertain. Ultimately, clarification here can only be achieved in individual cases, taking into account the respective product and its technical details.

8.2.5 Reallocation

The same delimitation questions arise in reallocation, that is, when previously valid weightings for asset classes or weightings due to market movements result in the initial weighting being restored at certain points in time.

In contrast to an initial event, asset management can often be excluded here by agreeing to fixed rules for reallocation and defining the timing and

[25] Bundesgerichtshofes in Zivilsachen, BGHZ 195, 126, X ZR 37/12; Oberlandesgericht Düsseldorf, NJW-RR 1073 (2016).

conditions of the adjustment so clearly that the provider does not have the conceptual discretion required for portfolio management. But then there is still the difficult question regarding investment advice or broking. The answer to this again lies in the individual structure based on the criteria presented.

8.3 Onboarding Obligations Under GWG

A key issue that has always played an important role for fintech firms is the client onboarding process, which is far more complex in the financial industry than in other sectors. We have already pointed out some regulatory aspects.

Financial service providers—irrespective of the type of license they operate under—are subject to the provisions of the GWG and, in particular, the obligations under Article 10 et seq. GWG (referred to as the know-your-client or KYC obligation). They are therefore obliged not only to collect the personal details of their clients but also to verify their identity, either at the post office counter or, as of recently, by video. This applies to the initiation of business transactions in general and to the opening of accounts and securities accounts in particular. This represents another significant hurdle and disruption to the onboarding process. The GWG also codifies organizational obligations. Fintech firms should therefore be careful when picking banking partners to ensure that they are open to working together on solutions that make onboarding as easy and efficient as possible, taking into account all legal and regulatory requirements.

8.3.1 Organizational Requirements

To prevent money laundering and terrorist financing, all obliged entities must have effective risk management systems in place, that is, perform and document risk analysis for their company and implement adequate internal safeguards. Article 6 GwG provides an overview of suitable safeguards. In accordance with Article 8 GwG, companies have a recording and retention obligation. The obliged entities are thus required to make complete copies of the identification documents or to record them digitally. For financial service providers licensed in accordance with the German Banking Act (KWG), it is also necessary to appoint a money laundering reporting officer (Article 7 GwG).

8.3.2 *General Due Diligence Requirements*

The general due diligence requirements (Article 10 GwG) include in particular:

- identifying the contracting party and, where applicable, the person acting on their behalf
- clarifying whether the contracting party is acting on behalf of a beneficial owner (and, if so, identifying the beneficial owner)
- obtaining and evaluating information on the purpose and intended nature of the business relationship
- establishing whether a politically exposed person (PEP) is involved (Article 1 (12) GwG contains a list)
- continuous monitoring of the business relationship, including transactions

With regard to identification, it should be noted that this must be done before the business relationship is established or the transaction is executed. This therefore has a direct impact on the onboarding process. In the case of natural persons, obliged entities are required to collect their name, place of birth, date of birth, nationality and address based on identity documents. To establish the identity of legal persons or partnerships, obliged entities are required to collect information about the company (i.e. name or trading name), the legal form, commercial register number, where applicable, the address of the registered office or principal place of business and the names of the members of its representative bodies or its legal representatives. Article 12 GwG specifies the documents to be used for identity verification purposes. The identity of natural persons can be verified on the basis of conventional ID documents. According to this, providers are required to verify the identity by examining the documents presented "physically" or using another procedure suitable for verifying identity in accordance with anti-money laundering laws and regulations. This also includes the video identification process and the Postident process. While the latter has long been recognized, BaFin has only recently specified the requirements for video identification. The BaFin Circular 3/2017, which is aimed at all financial institutions and service providers under its supervision, came into effect on 15 June 2017.[26]

Under the conditions specified there, the video identification procedure can now be used by all obliged entities under the Anti-Money Laundering Act. For example, the employees must receive special training and the person to be identified must give their explicit consent to the entire identification process being recorded at the beginning of the video identification. In addition, the image and sound quality of the communication must be sufficiently high to ensure unambiguous identification. The entire identification process shall be recorded and stored in an audiovisual format. Only end-to-end encrypted video chats are permitted for the necessary audiovisual communication. Services such as Skype are, therefore, not suitable for identification purposes. Against the background of these requirements, it is now possible to outsource video identification to several providers. At the same time, the requirements laid down in Article 17 GwG must be observed.

8.4 Conclusion of Contract

The conclusion of contracts online also requires closer examination.

Under MiFID2, a written framework agreement must be concluded with the client on paper or another durable medium, which sets out the essential rights and obligations of the investment firm and the client. Investment firms providing investment advice are only subject to this obligation where a periodic assessment of the suitability of the financial instruments or services recommended is performed. Accordingly, mere intermediaries are not subject to these obligations. However, portfolio managers and robo-advisors who have an ongoing relationship with their clients must comply with this requirement. As it is not typically possible to hand out a hard copy, a copy of the framework agreement must be sent to the client, for example, in pdf form.

The timing of the conclusion of the relevant contract also poses problems in this respect. As with any contract, digital contracts are also concluded through offer and acceptance. While clients routinely submit their offers online within the scope of robo activities, it is often forgotten that for the contract to come into effect, the offer needs to be accepted by

[26] See Bundesanstalt für Finanzdienstleistungsaufsicht (2017a).

the provider. This should be integrated into the onboarding process—for example, by using email.

It should also be borne in mind that digital business transactions are routinely concluded remotely, which entails special obligations for the company. Distance contracts are contracts where the company or a person acting on the company's behalf and the consumer exclusively use forms of distance communications to negotiate and conclude the contract. It is clear that this also applies to digital transactions. On the one hand, this means that a whole range of information must be made available to the consumer (cf. Article 246b of the Introductory Act to the Civil Code (EGBGB)). In the case of distance contracts, however, the business is obliged to provide the consumer with a confirmation of the contract, on a durable medium, in which the content of the contract is set out, and to do so within a reasonable period of time after having concluded the contract, but at the latest before the performance of the service. Depending on what has been said above about regulatory law, this gives rise to an obligation to provide the client with the agreement.

Finally, the consumer's right to cancel must be taken into account in distance contracts. If the conditions of a distance contract are satisfied, the consumer will have the right to cancel the distance contract. The cancellation period is two weeks. However, in order for the two-week period to start to run properly, its start is subject to a number of conditions, which are difficult to meet in practice. This includes the special information concerning the right of cancellation and the provision of information in accordance with Article 246b EGBGB. The business must provide the consumer with the information in written or electronic form at the latest by the point at which the obligations under the contract have been met in full. If this does not happen, the cancellation period does not begin to run—resulting in an infinite cancellation right within the scope of the financial services.

8.5 Data Management

Robo-advisors collect and process a variety of sensitive client data. In this respect, they are subject to the strict requirements set out in the General Data Protection Regulation (GDPR) and the German Federal Data Protection Act (BDSG). In addition to various organizational requirements, robo-advisors have to take into account, in particular, the strict requirements for the lawfulness of data processing and the far-reaching information obligations that must be observed during onboarding.

However, there are two aspects which need to be highlighted in relation to fintech firms and robo-advisors.

8.5.1 *Profiling*

GDPR specifically focuses on so-called profiling. In a nutshell, profiling means the automated creation of profiles. In connection with web-based systems, this mainly involves the creation of user profiles on the basis of automatically collected personal data, taking personal aspects into account. This is precisely what robo-advisors usually do for clients as part of their investment proposals. Providers are not only required to inform their clients about profiling, but they also have to meet specific requirements set out in Article 22 GDPR with respect to the lawfulness of this type of data processing.

Profiling is always prohibited if the processing of personal data is solely automated, which produces legal effects concerning the data subject or similarly significantly affects him or her. It should be noted that the restriction contained in Article 22 (1) GDPR relates to decisions based "solely" on automated processing of personal data. In addition, Article 22 (2) GDPR sets out the circumstances under which profiling is permitted after all.

This may be the case, for example, if it is necessary for entering into, or performance of, a contract between the data subject and a data controller. This refers to cases in which the conclusion or performance of the contract reflects the wishes of the data subject, who thus does not consider fully automated processing to be an infringement of his rights. This exception should, therefore, routinely work for robo-advisors with regard to the provision of financial services. It is also important to note that under recital 71, profiling and any other form of automated data processing should not concern a child. This restriction must be observed at all times.

In addition, there are special rights and obligations, which must be observed, including the right to contest the decision, to express his or her point of view and to obtain human intervention.

8.5.2 *Use of Cloud Services*

Robo-advisors regularly use cloud services to store data or to meet certain requirements (such as retention requirements). The use of cloud computing by credit and financial services institutions may, in individual cases, involve outsourcing that is material to the execution of financial services or any of the institution's other usual services within the meaning of Article 25b KWG. The isolated purchase of software is generally to be classified as another purchase, and it thus does not constitute outsourcing. However, this does not apply to software which is used to identify, assess, control, monitor and communicate risks or which is essential for the performance of institution-specific tasks. Support services provided for this type of software are classified as outsourcing.[27] Furthermore, the operation of the software by an external third party is also classified as outsourcing.

As cloud computing software does not work locally on the server of the robo-advisor but is operated routinely as a cloud-based service, it can be assumed that this—provided it concerns part of the advice, broking or management services—qualifies as outsourcing within the meaning of Article 25b KWG.

In this respect, the qualifying criteria for use of external services have been significantly tightened by the new version of Minimum Requirements for Risk Management (MaRisk) 2017. While under the old rules the outsourcing of business activities only took place if an institution engaged another company (outsourcing company) to perform an activity or function (service) that was material to the activity of the institution in the long term or at least for a longer period of time, under MaRisk 2017, the criterion of materiality is no longer applied to the question of whether cloud computing qualifies as outsourcing.

Accordingly, the use of cloud services must always be assessed to set up an outsourcing process where appropriate. The first pool audits as part of the regulatory review of large cloud providers took place in 2018.

[27] Helpful for case-by-case assessment: BaFin's Guidance Notice on outsourcing to cloud providers, see Bundesanstalt für Finanzdienstleistungsaufsicht (2018).

8.6 Conclusion

The regulatory requirements are continually evolving in line with the developing technical infrastructure. Humans and machines will continue converging. Standardized technical processes assist clients in their decision-making and help to eliminate human errors to a large extent. Regulators will therefore continue to supervise the robo-advisor market and make the solutions more efficient and forward-looking. MiFID 3 will be the next major regulatory amendment in the long term.

References

Assmann, Heinz-Dieter, Rolf A. Schütze, and Petra Buck-Heeb. 2015. *Handbuch des Kapitalanlagerechts*. 4th ed. Munich: C.H.BECK.

Boos, Karl-Heinz, Reinfrid Fisher, and Hermann Schulte-Mattler. 2016. *KWG, CRR VO–Kommentar zu Kreditwesengesetz, VO (EU) Nr 575/2013*. 5th ed. Munich: C.H.BECK.

Bundesanstalt für Finanzdienstleistungsaufsicht. 2016. *Robo-Advice und AutoTrading–Plattformen zur automatisierten Anlageberatung und automatischem Trading*. Last Accessed 20 March 2020. https://www.bafin.de/DE/Aufsicht/FinTech/Anlageberatung/anlageberatung_node.html.

Bundesanstalt für Finanzdienstleistungsaufsicht. 2017. *Circular 3/2017 (GW) – Video Identification Procedures*. Last Accessed 20 March 2020. https://www.bafin.de/SharedDocs/Veroeffentlichungen/DE/Rundschreiben/2017/rs_1703_gw_videoident.html.

Bundesanstalt für Finanzdienstleistungsaufsicht. 2017. *Merkblatt Anlagevermittlung*. Last Accessed 20 March 2020. https://www.bafin.de/SharedDocs/Veroeffentlichungen/DE/Merkblatt/mb_091204_tatbestand_anlagevermittlung.html.

Bundesanstalt für Finanzdienstleistungsaufsicht. 2018. *Merkblatt – Orientierungshilfe zu Auslagerungen an Cloud-Anbieter*. Last Accessed 20 March 2020. https://www.bafin.de/SharedDocs/Downloads/DE/Merkblatt/BA/dl_181108_orientierungshilfe_zu_auslagerungen_an_cloud_anbieter_ba.html.

Bundesanstalt für Finanzdienstleistungsaufsicht. 2020. *Gemeinsames Informationsblatt der Bundesanstalt für Finanzdienstleistungsaufsicht und der Deutschen Bundesbank zum Tatbestand der Anlageberatung*. Last Accessed 20 March 2020. https://www.bafin.de/DE/Aufsicht/FinTech/fintech_node.html.

Löhr, Andrea, and Daniela Rothe. 2020. *Geeignetheitserklärung: Wichtiges Dokument für Verbraucher*. Last Accessed 20 March 2020. https://www.bafin.de/SharedDocs/Veroeffentlichungen/DE/Fachartikel/2018/fa_bj_1809_Geeignetheitserklaerung.html.

Oppenheim, Robert, and Christian Lange-Hausstein. 2016. "Robo Advisor – Anforderungen an die digitale Kapitalanlage und Vermoegensverwaltung." *WM Zeitschrift für Wirtschafts- und Bankrecht*, 1966–1973.

Schwark, Eberhard, and Daniel Zimmer. 2010. *Schwark/Zimmer–Kapitalmarktrechts-Kommentar*. 4th ed. Munich: C.H.BECK.

Tertilt, Michael, and Peter Scholz. 2018. "To Advise, or Not to Advise – How Robo- Advisors Evaluate the Risk Preferences of Private Investors." *The Journal of Wealth Management* 21 (2): 70–84.

PART III

Case Studies of Robo-Advisory

CHAPTER 9

(Re-)Launching a Robo-Advisor as a Bank

Theodor Schabicki, Yvonne Quint, and Soeren Schroeder

Contents

This chapter is also published as white paper "Robo-Advisory in the Securities Market" by BearingPoint.

T. Schabicki (✉) • Y. Quint
BearingPoint GmbH, Frankfurt am Main, Germany
e-mail: theodor.schabicki@bearingpoint.com; yvonne.quint@bearingpoint.com

S. Schroeder
BearingPoint GmbH, Hamburg, Germany
e-mail: soeren.schroeder@bearingpoint.com

© The Author(s) 2021

P. Scholz (ed.), *Robo-Advisory*, Palgrave Studies in Financial Services Technology, https://doi.org/10.1007/978-3-030-40818-3_9

9.1 Introduction

Robo-advisors in today's known form appeared in the UK in 2008 for the first time and provided customers with a simple and inexpensive way to invest money. Since their market entry, robo-advisors have developed rapidly, and by now affect the financial sector's securities business significantly, more specifically:

- In 2018, of the 800,000 new DIY investment accounts opened in the year to the end of September, one in three were opened with one of the UK's main robo-advisors, including Nutmeg and Moneyfarm (Source: FT, November 2018).
- The number of DIY investment accounts—including robo and platform customers—rose to 4.8 m over the year to September, an increase of 22%.
- The volume of assets in the DIY investing market, where customers pick and choose investments without the help of a financial adviser, grew to £224 bn over the 12 months to September and a 15.4% increase over the year.

In order to participate in the market growth and to be able to remain competitive in the securities business, banks incrementally cooperate with fintechs to ensure the implementation of their own robo-advisors. However, for a successful implementation or further development, some key considerations must be made.

This chapter addresses all decision-makers in the securities business from banks to investment service providers, and addresses the following two questions:

- Firstly, how should robo-advisors be introduced for the first time?
- Secondly, how could the existing robo-advisors' success be evaluated?

9.2 Background

For more than 10 years, the stock markets have shown an almost constant growth curve. The sustained stock market development, combined with a lack of investment alternatives, has led many private investors to invest their money in securities. Over this period of economic expansion, the quality of products and services offered by banks was of low importance due to a lack of alternatives. Driven by continuously increasing share prices, banks were actively giving generic investment advice, paired with traditionally expensive products, leading to steadily increasing securities revenues. Banks were thus able to continuously increase their securities revenue, even with average investment advice and relatively expensive products .

However, more and more innovative and agile fintechs have entered the securities business and are challenging established financial institutions with their own digital product offerings. Robo-advisory has become increasingly prominent and thus been mentioned as an investment product in various news and media channels.

The formula for success concerning robo-advisors focuses on transparent, cost-effective—and above all—digital product offerings. Robo-advisors offer their customers intuitive and easy-to-understand digital asset management by providing an exclusively digital advisory process. The portfolio is derived by using a risk profiling process, and is usually mapped by the robo-advisor via a cost-effective ETF portfolio. This approach contrasts considerably with the banks' relatively opaque, expensive and non-digital securities products.

In order to compete effectively with fintechs, banks increasingly pursue cooperation agreements with fintechs and offer their own customers robo-advisors in the form of white-label solutions. The benefit of this approach is that white-label solutions can be implemented much faster than in-house developments as they are designed upon the existing expertise of fintechs. Additionally, many fintechs have also expanded their business model to include cooperation agreements with banks, with some focusing exclusively on this.

Offering a robo-advisor itself is no sure formula for success for banks. The robo-advisors offered by banks in cooperation with fintechs differ only slightly from sole fintech offerings, and are significantly more expensive than fintechs on average, positioned as niche products without a strategy in place. Consequently, the probability of a sustainable product success rate is low. In order to maximise the chances of success, banks must deal with the individual institutional challenges of robo-advisor implementation and create the strategic and/or organisational framework for the digitalisation of securities advising.

9.3 Key Challenges and Considerations

Many banks still process most of their securities business by means of traditional advice from securities specialists. Digital channels (in terms of self-service offerings) often only exist for the non-advisory aspect of business. Implementing a robo-advisor, the original advising business concerning securities can be digitalised for the first time. The main challenges of such an implementation project lie in the following context (Fig. 9.1):

The key challenge of introducing a robo-advisor is to develop a clear target vision for the securities business. The robo-advisor does not correspond the introduction of a new product, but rather to the beginning of the

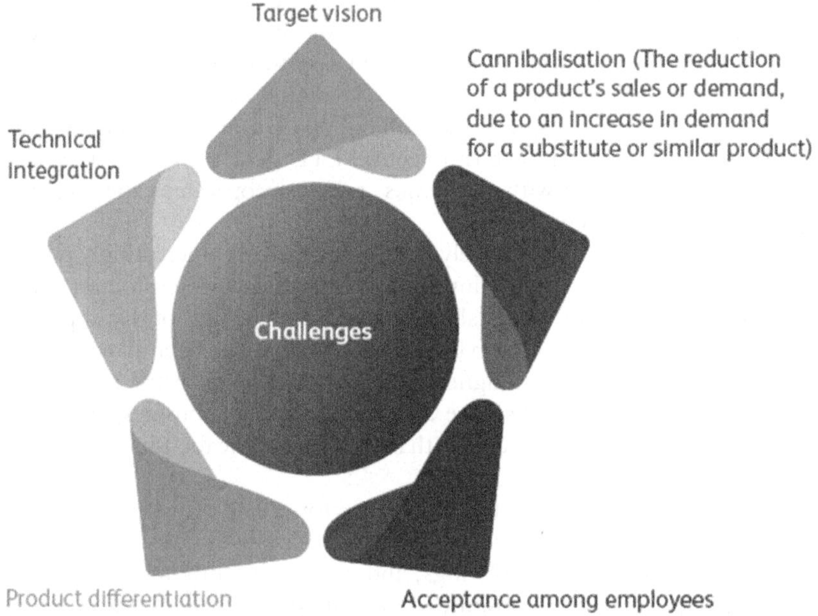

Fig. 9.1 Challenges of implementing/introducing a robo-advisor

digitalisation of the advising business and thereby also leads to a profound adjustment of the business model. This means the bank must deal with the following questions: Which distribution types (execution only, investment advice, asset management) should be offered in the future, how are the distribution channels to be linked, and how can coexistence between man and machine take place? Only if the target vision is defined precisely, the role of the robo-advisor (and therefore the securities advisor) will become clear.

The introduction of a robo-advisor is inevitably paralleled by fears in management of possible cannibalisation effects. On the one hand, high earnings targets are set for the securities business to compensate the low income from interests. On the other hand, the margin of a robo-advisor is significantly lower than the margin of traditional securities products. The response of sales management is to integrate the robo-advisor into the marketing control with too much care, and to carry out external marketing in a too controlled manner. Overcoming these cannibalisation fears in management is critical to success. A clear and well-communicated vision and support from top management helps to address this challenge and to embed the robo-advisor in the securities business sustainably.

More significant than the management's fears of cannibalisation are the security advisors' existential fears that arise due to the introduction of a robo-advisor and the associated lack of acceptance of the solution. The advisors become sceptical that digital tools such as robo-advisors will question their position in the securities business value chain. As many clients maintain long-term and trusting relationships with their advisors, it is of utmost importance to convince the advisors that robo-advisors bring additional value. Securities advising is not possible without the advisors, not only in current terms but in the future too. However, this role must be reconsidered in regard to the relationship between man and machine—undoubtedly a major challenge for banks and asset managers. A possible scenario for the future is for advisors to develop into solely relationship managers, while the robo-advisor covers all the technical aspects.

Another challenge is to clearly differentiate one's own product from the competition. Many robo-advisors from banks differ very little from the offers of fintechs for instance, profiling, asset allocation and financial products are all typically similar. The only perceived difference for the customer is price. Much to the displeasure of banks, in order to achieve justifiable margins they must offer their robo-advisory services at a much higher price on average than most fintechs. Consequently, customers prefer to give their money to the original provider (fintech) instead of a replica developed by a bank. Therefore, it is crucial for banks to differentiate their own robo-advisors from fintech products in a way that adds value and is noticeable to the end client.

The introduction of a robo-advisor is accompanied by technical challenges. For instance, in a case where the robo-advisor is implemented with an external partner (instead of in-house development), the degree of integration within the bank's existing IT systems must be determined clearly. Although a deep integration leads to a significantly optimised customer experience and more efficient processes, it also results in considerably longer project lifetimes, costs, risks and a higher dependency on the supplier.

9.4 Implementation of a Robo-Advisor

Banks have recognised the fundamental necessity of digitalising securities advising in regard to a self-service offering and are starting the digitalisation process with the introduction of robo-advisory services for asset management. Since the implementation should happen as quickly as possible to avoid significant IT impact, an in-house development or extension of traditional asset management by a digital channel is often not an option. The fintech market is broad, considering the fact that many firms already successfully offer white-label robo-advisory solutions in the industry.

However, it is often assumed that robo-advisory services can be introduced as a new product within a lean implementation project. This may apply to the provision and connection of the technical platform, but not to the definition of the necessary strategic and organisational framework required.

For ensuring the sustainable success of a robo-advisor, it is crucial to define the strategic framework conditions first and commence the robo-advisory implementation in that context.

9.4.1 Strategic Framework and Parameters

Historically, banks' securities business is defined by complex advising processes, extensive financial product portfolios and a rigorous sales management focused on network marketing. In this network, the advisor leverages an outstanding position as their advisory services have a significant influence on the success and quality of the securities business. In order to successfully place a robo-advisor within the market, the robo-advisor should not be regarded as a product but as an enabler to the securities business. The strategic framework parameters must be defined prior to the implementation. Three crucial questions must be addressed during that stage:

- Which forms of distribution will be used to implement the securities business in the future?
- How should analogue and digital distribution channels interact?
- What should the future role of the (human) securities advisor look like?

Regarding securities, banks, as a basis, offer three different forms of distribution; execution only, investment advice, and asset management. Prior to the start of the digitalisation of the securities business, the bank needs to make two fundamental decisions. Firstly, it is necessary to determine which distribution form the bank intends to use in the future for its securities business. This is crucial, since the complexity of such a project is greatly increased by the digitalisation of the investment advice service, in terms of a self-service offering by comparison to asset management. The fundamental question of whether investment advice should be offered in the future, or whether the entire advising business should be subsumed by asset management, is especially interesting. The latter will enable the securities business to be implemented much more efficiently, as it will centralise the investment decision, generate returns free of subsidies, and create less documentation obligations than the investment advice process.

Secondly, it is necessary to determine how the possible duplication of products in asset management will be dealt with once the robo-advisor has been implemented. Most banks offer two-asset management services post the introduction of a robo-advisor—traditional asset management and digital asset management, provided by the newly implemented robo-advisor. If the traditional form of asset management is to be continued, a clear product differentiation is essential. If asset management is to be consolidated in the future, a consolidation strategy should be developed at the time when the robo-advisor is introduced.

In addition to the definition of future sales forms, the interaction between the traditional and the digital sales channel must be defined. The area of conflict ranges from two independently offered sales channels, to an omnichannel approach, in which the customer can switch between the defined channels as and when needed, that is, advisory begins online and is concluded together with the securities advisor. However, this significantly increases the requirements for the technical integration of the robo-advisor into the core banking systems and requires significant adjustments to the pre-existing sales process to ensure compatibility.

The implementation of a robo-advisor will significantly change the securities advisor's profile. So far, the advisor has been responsible for making asset-allocation decisions and buying/selling securities together with a customer. In the future, the securities advisor will be released from these tasks, as the robo-advisor handles the entirety of the technical side, from risk profiling and the derivation of asset allocation, to the purchase/sale of financial products. Consequently, the role of the advisers must be redefined. This is important to prevent a possible threat to employment, and thus the rejection of the robo-advisor implementation plans by the advisers at an early stage.

9.4.2 Robo-Advisory Specification

Once the strategic framework conditions have been set, the specification of the robo-advisor in terms of target group, financial ratios, sales management, consulting process and integration depth is carried out (Fig. 9.2).

When determining the target group, age, IT affinity, financial circumstances and expected service level of the customers should specifically be considered. The target group for the robo-advisory service should be distinguished from the target groups of other securities products to ensure an optimal customer approach.

The financial ratios are made up of the expectations regarding the definition of volume and price. Volume planning should be based on the overall portfolio volume development for the securities business, as well as the relative allocation of the individual sales forms (execution only, investment advice and asset management). For the definition of volume of robo-advisors, the integration with stationary sales, as well as the scope of the marketing measures, are critical to success. The price of the robo-advisor consists of the financial product costs and the service fee for the financial portfolio management, including securities account management. Externally, the competitors' prices and the prices of comparable institutions with similar target groups are to be evaluated. Internally, the price needs

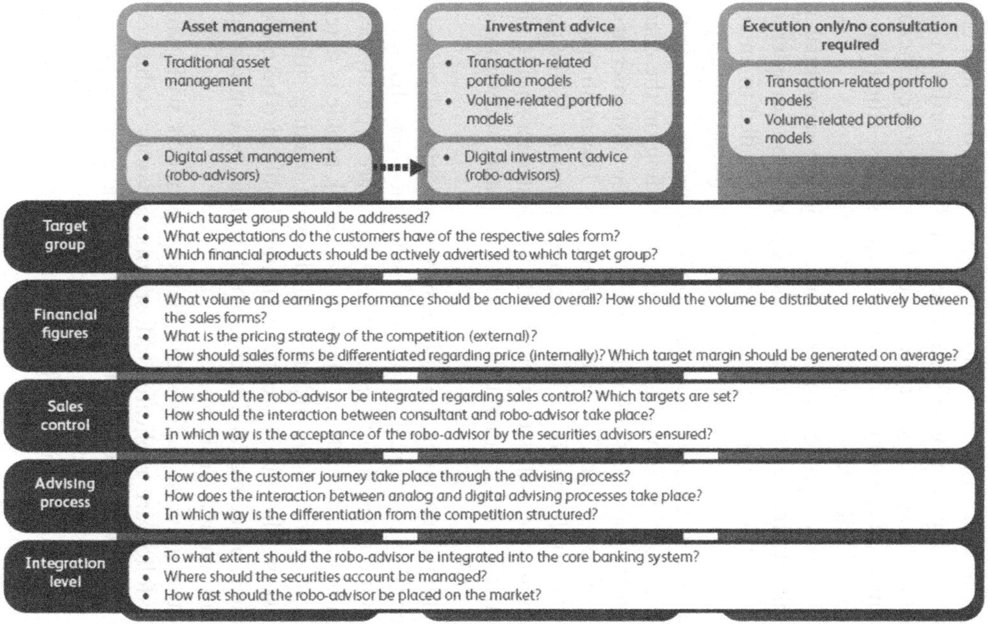

Fig. 9.2 Dimensions for specification of the robo-advisor in the context of the overall securities strategy

to be considered in the context of the existing financial product portfolio. Due to the immense pressure of competition, a pricing approach in line with the market often results in a lower target margin compared to classic securities products.

In contrast to fintechs, the relationship between advisor and customer is based on a trusting advisor–customer relationship. In order to integrate the robo-advisor into such a structure in a useful way, the securities advisors must be incentivised regarding the robo-advisor's sales—the robo-advisor has to be a part of the sales control.

The robo-advisory process mainly consists of five sequential process steps. Both distribution and regulatory aspects must be considered when specifying the individual process steps. It is recommended to amalgamate the advisory process of the robo-advisor manually to the advisory process of traditional asset management advice from both perspectives, in order to avoid different securities advising processes, depending on the channel.

The onboarding process involves identifying the customer according to the KYC (Know-Your-Customer) procedures. The main question in this respect is who the customer's contractual partner is. As the customer relationship is an essential asset of the bank, the customer should close the financial portfolio contract directly with the bank, and not with the robo-advisor.

Risk profiling determines the risk-bearing capacity and propensity of the customer. This is normally processed using questions that are easy to understand. Questions about behavioural finance are typically becoming increasingly popular. Therefore, the customer's risk profile is determined by "softer" questions about hypothetical decision-making situations. In addition to risk profiling, the potential customer should also be asked about their investment objectives and needs, which should be considered in the context of personal risk profiling.

The product universe defines the financial products with which the different portfolios are replicated. Most robo-advisors exclusively offer ETFs due to the favourable cost structure of around 20 basis points. Only a few robo-advisors additionally mix the portfolios with actively managed funds or individual securities. The product universe should always be defined by the bank.

A technical aspect is the depth of integration of the robo-advisor into the bank's IT architecture. The depth of integration describes the degree of integration of the new application within the core banking system, or within other relevant systems. The extent of the integration defines the area of conflict between the scope of functions and the implementation period. The deeper the integration, the higher the functional scope, but the more complex and time-consuming the implementation. Here, the main question is whether the securities accounts should be managed by the bank or an external custodian bank.

9.4.3 Evaluation Criteria for Provider Selection

If a white-label solution is used, the criteria of the strategic framework and the specification of the robo-advisor are defined for the evaluation and the subsequent selection of the provider. The criteria should be grouped and weighed according to their relevance. A possible grouping can be achieved by looking into the following dimensions (Fig. 9.3):

9.4.3.1 Functional Specification

The basis for the evaluation of the functionality is the defined advising process. The extent to which the defined requirements are met by possible providers should be evaluated separately for each process step. The key factor here is the degree of individualisation of the advisory process. For instance, is it possible to use an individual risk profiling approach, or is it necessary to adopt the provider's profiling approach? A high degree of individualisation of the solution enables the bank to use the new digital advising process in line with the existing analogue process and allows flexible adjustments for the future. In order to interlock robo-advisors and stationary sales, it must be ensured that the provider also offers special consultation access points which allow the securities advisors and customers to use the robo-advisors together.

9.4.3.2 Technical Specification

Considering technical aspects, such as data protection and the integration of the solution into the bank's existing systems, is vital to the success of the technical specification. In regard to data protection, it is necessary to determine on which IT infrastructure the robo-advisor service will be operating. Both the location of the data storage and the location of the data access are relevant and must be adaptable to the needs and requirements of the bank. In general, banks expect data storage to be in the UK, and access only available from locations within Europe. Certifications or regular independent external audits can certify the provider's compliance

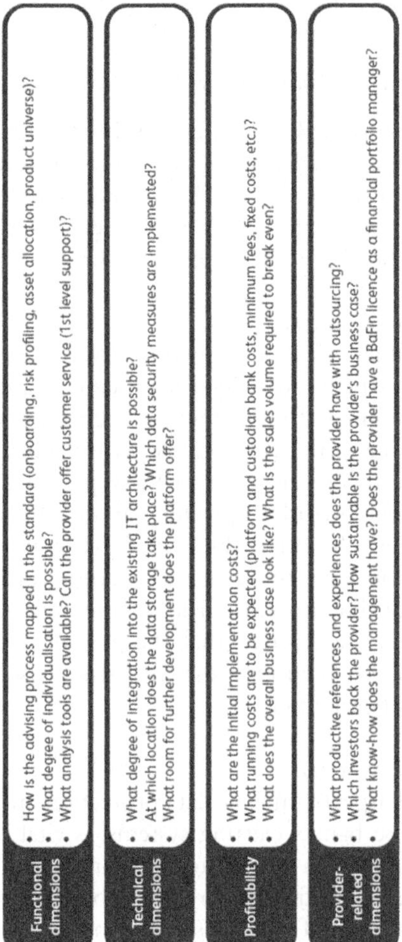

Fig. 9.3 Evaluation dimensions for provider selection

of security standards. In addition to security, the legally compliant handling of customer data must also be ensured.

When it comes to the technical integration of the solution, experience—or even existing interfaces between the provider and the bank's core banking system—banks must ensure diverse advantages during an implementation. An open architecture in terms of platform banking and the provision of programming interfaces and so-called APIs, gives third parties a simple and standardised connection. Such a technological basis offers the possibility of extending the service through third-party service providers.

9.4.3.3 Profitability

The challenge in evaluating profitability is that providers often have different cost structures and are thus not directly comparable with each other. In addition to the usual platform fees as a percentage of the assets under management, providers sometimes charge minimum fees and fixed service and/or maintenance fees. However, in other cases the implementation costs differ considerably, depending on the provider.

In order to establish comparability between providers, the evaluation must be carried out based on a business case. For the business case, the bank makes projections about the development of the custody account volume and the price for the end customer. Once these projections have been made, the predicted revenues can be compared with the costs of the providers. In the case that securities accounts are held by an external custodian bank, these costs must also be considered in the projection. Since minimum fees for individual providers can significantly influence the business case, the case of low volumes and various scenarios for volume development need to be calculated.

9.4.3.4 Provider-Related Specification

Lastly, the provider analysis will need to take place. The core objective of this analysis is to evaluate how competitive the provider is and how efficiently a potential implementation project can be carried out together with the provider.

The successful implementation of reference projects is probably the best indicator for a successful implementation. In this case, the provider is already familiar with the typical challenges, has already implemented best practices, and can demonstrate sufficient support for the sales strategy.

Other criteria include the accreditation of a financial portfolio manager, a proven history of years of successful market presence, financial stability and sustainability, and financially strong investors. Furthermore, the provider's cooperation partners also have an impact on the business model. For example, in the case of an external custodian bank being commissioned to manage the client portfolios, it is advantageous if the provider has already proposed or established a relationship with a cooperation partner.

9.4.4 *Implementation of the Robo-Advisor*

The considerations for strategic integration, the defined requirements for the supplier, and the evaluation grid, result in a framework whereby the market for robo-advisors is viable to explore. This heterogeneous market can be divided into three segments.

1. Some of the robo-advisory providers only operate within the B2B market. These companies support banks and asset managers in developing and providing a robo-advisory service. As software developers, they do not directly make contact with the investor as an end customer.
2. Providers are operating in both the B2C and B2B markets. Many of these companies have originated their business model in the B2C market and are now expanding by offering white-label solutions for B2B clients.
3. The market for only B2C providers consists of fintechs and digital solutions from established financial institutions. The development of these solutions is partly based on the state of the B2B market.

Provider selection is based on the previously defined evaluation criteria and should be carried out with a systematic process. The result of the multi-stage process is the decision for an appropriate cooperation partner for the implementation of a robo-advisor (Fig. 9.4).

9.5 Evaluation of Existing Robo-Advisors

Many banks have already implemented their own robo-advisors as white-label or in-house developments. However, the success of these robo-advisors is by no means guaranteed. There are already first-hand examples

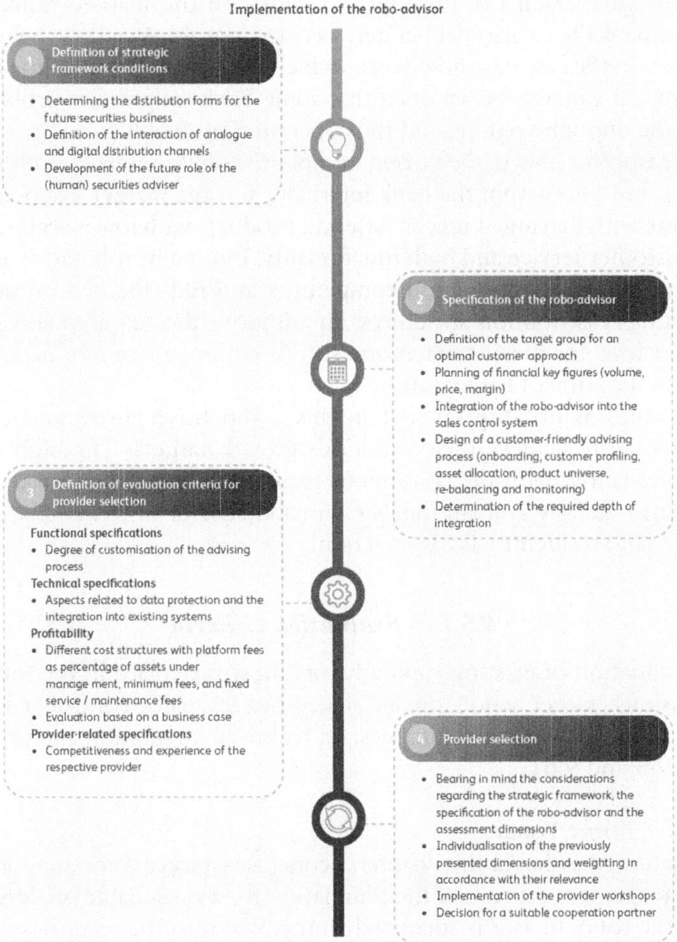

Fig. 9.4 Introduction of a robo-advisor

of failed partnerships of fintechs and banks on the market. Some commercial banks have also deliberately decided not to offer their own robo-advisors. In the case of robo-advisors which have been placed on the market by banks, it can also be assumed that some banks might face problems in acquiring enough volume, and thus operating profitably.

A reason for this is the strong competitive environment, both on the market, but also within the bank internally. On the market, robo-advisors compete with dynamic fintechs, offering products with low margins, excellent customer service and high functionality. Internally, robo-advisors have to compete with high-margin competitors and old (focused on network marketing) distribution structures. In addition, the required service and comfort level for the customers and securities advisors cannot be achieved due to suboptimal IT integration.

It comes as no surprise that in this competitive environment robo-advisors fail to achieve their ambitious growth targets. The high market dynamics paired with the continuous technological progress make it necessary to regularly evaluate one's own robo-advisor in terms of securities strategy and to identify fields of action.

9.5.1 Evaluation Criteria

The evaluation of existing robo-advisors must take place in the following dimensions: target vision, financial key figures, sales control, customer usage and satisfaction, product design, technical integration and providers (Figs. 9.5 and 9.6).

9.5.1.1 Target Vision

As mentioned earlier in this chapter, a conclusive target vision supported by top management represents the foundation for a sustainable process. Only when the robo-advisor is adequately integrated into the securities strategy of the bank is it possible to enhance further development. Additionally, the sustainability of the overall target vision and the specific consideration of the robo-advisor should be evaluated. On the other hand, it is also important to validate the communication and transparency of the proposed strategy within the organisation. The target vision needs to define the framework conditions for the specific design of the robo-advisor.

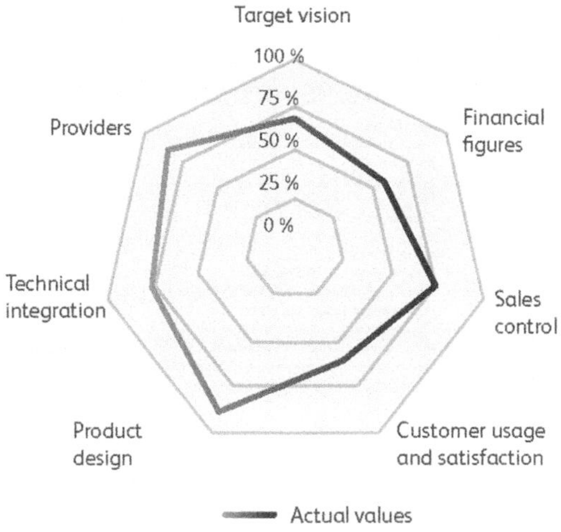

Fig. 9.5 Evaluation criteria for existing robo-advisors

9.5.1.2 Financial Figures

In the long term, the financial figures are the key measures for evaluating the robo-advisor's success. This includes key figures regarding the achieved volume, the achieved revenues and margins, but also the costs for the provision of the service itself. The evaluation is carried out as a comparison, with predicted values, or as a benchmark against the market. The financial ratios of the robo-advisor should also be compared with the rest of the bank's securities products.

9.5.1.3 Sales Control

To ensure the sustainable acquisition of custody account volumes, the robo-advisor must be carefully considered within sales management. This includes the overall planning for the securities business, as well as the specific sales targets. Both nominal targets and the relative relationship to other control parameters must be evaluated. In addition to the control parameters, the way the targets are supported by marketing and training measures must also be evaluated. The acceptance of the robo-advisor by the securities advisors is also paramount in this dimension. In banks with an intensive customer–advisor relationship, it is critical to success to ensure that the advisors regard the robo-advisor as a value-adding activity, and thus recommend the service to customers alongside traditional

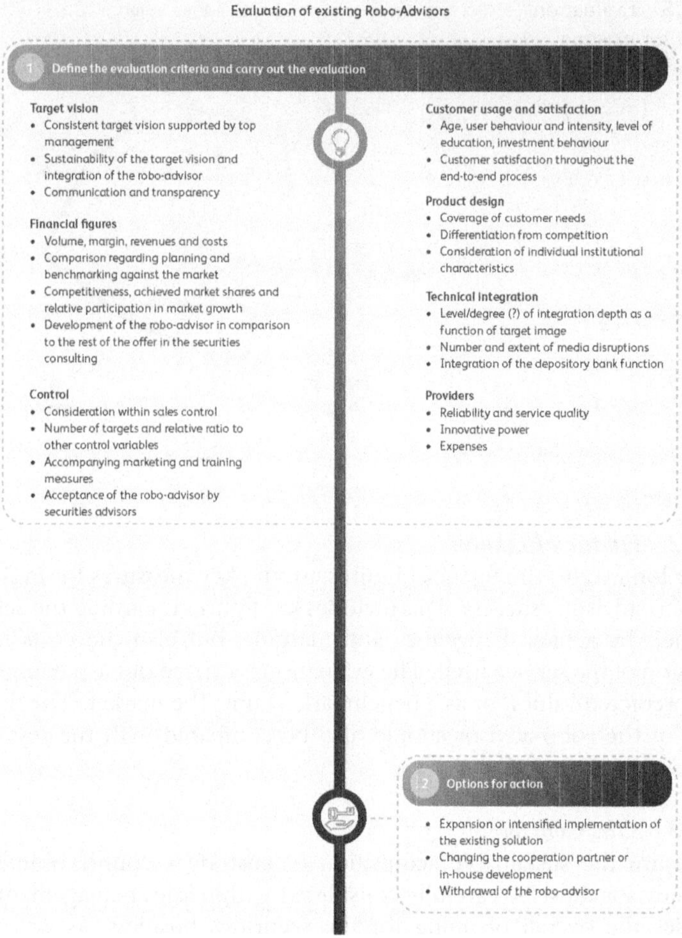

Fig. 9.6 Evaluation of existing robo-advisors

product offerings, bearing in mind not to offer comparable or overlapping solutions.

9.5.1.4 Customer Usage and Satisfaction

Two preliminary questions to evaluate the success of a robo-advisor detail the customers general usage and how satisfied they are with the solution. To answer the first question, information concerning customer characteristics must be evaluated. The information provided includes age, user behaviour, usage intensity, education level, investment behavior (amount and time) as well as other aspects. In principle, the aim of gathering this information is to assess which group of customers use the robo-advisor in order to address comparable customers in a targeted way with regard to the product, or to align sales control appropriately. To answer the second question, the users of the product must be questioned directly in an online survey. In this context, customer satisfaction should be explored concerning the entire end-to-end process. This includes the initial approach to the product, the advising process, the closure, the reporting and the interaction between the securities advisor and robo-advisor. These results enable valuable conclusions to be drawn for further development of the solution.

9.5.1.5 Product Design

In the context of the product design, the aim is to evaluate the degree to which the robo-advisor meets the customer needs and how the solution differentiates itself from the competition. The customer survey can be used to gain insights as to where in the process the robo-advisor should be optimised, in order to reflect the customer needs. Furthermore, the robo-advisor should be compared with competing products via benchmarking processes. In particular, it is important to assess the degree in which the robo-advisor differentiates itself from the competition, and which features are specific to the institution.

9.5.1.6 Technical Integration

The degree of technical integration required is derived from the target vision. However, for long-term success, it is important that the robo-advisor is connected to the existing system landscape. This dimension evaluates how well the robo-advisor is integrated into the bank's existing systems. In this context, it is essential to establish to what extent media disruptions occur for customers (e.g. lack of single sign-on or missing assumptions of existing risk profiles). Another evaluation criterion is the integration of the custodian bank function, which is currently often outsourced to external custodian banks, creating additional complexity for the customer in terms of a second point of contact.

9.5.1.7 *Providers*

In many robo-advisory solutions, external providers are used for the financial portfolio management and in part for the custodian bank function too. Reliability, innovation and service quality are essential evaluation criteria for successful cooperation with the provider. This is essential if the robo-advisor's frontend is to be used for other securities products in the future (e.g. investment advice), the further development of the securities business depends significantly on the strength of the provider.

9.5.2 *Options for Action*

The evaluation result is the basis for a purposeful optimisation of the existing robo-advisor solution. Three general options can be derived from the spectrum of possible actions, which can be implemented individually for each institution:

- Ongoing development of the existing solution
- Fundamental change regarding the existing solution
- Shutdown of the robo-advisor

In the presence of constantly changing customer requirements and market changes, further development of the existing solution is inevitable. By deepening the integration into the bank's systems, for example, the customer experience can be significantly optimised by building on already-existing customer data. When expanding the existing robo-advisor, the market-related developments, the institute-specific framework conditions and the innovation potential of the white-label provider must be taken into account. It is crucial that the robo-advisor is meaningfully integrated into the bank's securities strategy.

A fundamental modification of the existing robo-advisor solution must be considered if the existing white-label provider cannot implement essential requirements, or if the innovation potential for an expansion is missing. In these cases, the white-label provider could be replaced, or a separate solution could be developed. However, it should be established how the existing customer relationships should be handled. Customer custody accounts are often held with an external custodian bank and cannot be transferred without contractual adjustments.

The withdrawal without any alternative option from the robo-advisor business is an action up for questioning: The combination of demographically induced shifts in customer behavior and the financial maturity of the digital natives' demand for digital solutions in the securities business. In the short term, a withdrawal might improve financial figures. However, this is at the cost of a weakening of competitiveness in the securities business. Digital asset management in the form of a robo-advisor can be regarded as a fundamental step in ensuring a sustainable digitalisation of the securities business.

9.6 Future Trends

From digital securities advising to digital asset management—technology trends, which can be observed in other areas of digital banking, are also finding their way into asset management. In this context already-known innovations are being used:

- Voice Banking & AI—contactless, voice-based banking, which continuously improves the understanding, interpretation and prediction of customers' needs by using Artificial Intelligence (AI) in a self-learning manner.
- Digital Identities—central management, allocation and safe custody of uniquely assigned identities and corresponding collateral information.
- Blockchain infrastructures—substitutes for traditional and securities accounts as well as being the basis for trustworthy, highly efficient (business) transactions, especially in the area of digital identity or legally secure signatures.
- Open Banking APIs—flexible, rapidly deployable interface technologies that greatly simplify and accelerate interaction with other market participants or customers.
- Marketplaces (SaaS)—establishment and operation of a regulatory and trust-based framework to better serve "supply and demand".

The future development in the area of digital securities advising will probably show two different characteristics:

- On the one hand, there will be massive consolidation pressure among B2C providers, as customer acquisition costs (CAC) and customer lifetime value (CLV) can only be turned into a positive business model in larger organisations, by benefitting from the effect of economies of scale. In this context, the innovation potential of the resulting organisations, which will be exposed to similar competitive pressure as existing competitors in the non-digital market, appears questionable.
- On the other hand, we believe that in the B2B segment, models will prevail, allowing customers easier access to software components and business processes. Due to the strategic advantage of platform-based structures, it is expected that for the area of digital asset management, software-as-a-service (SaaS) marketplaces will emerge, enabling corporate banking customers to implement a business process innovation or add new digital components to an existing process with a few simple steps.

9.6.1 Digitalisation Requires Individualisation

Today's innovation still looks as follows: for each stage of digitalisation, the best solution for achieving pre-defined goals must be worked out from a variety of possible providers (internal and external) at first. In larger IT projects, various integration procedures and connections to different systems must be developed in order to integrate the corresponding components. These usually require an excessive amount of time, money and resources.

9.6.2 Marketplaces and Ecosystems

The individualisation approach is contradicted by the logic of an SaaS marketplace, connecting different providers of subcomponents in advance via standardised interfaces ("APIs") and making them available to the respective users individually on demand. In this context, we shape the term "survival of the API-est", as we believe that in the future only those market participants will assume significant ecosystem roles and master their APIs. The API management systems used for this purpose control and regulate the connectivity between the providers and their services, with the corresponding consumers and customers. Therefore, from our experience, the success of a marketplace operation depends essentially on the effective use of APIs. The marketplace operator itself is usually only a provider of SaaS solutions in certain sub-segments—rather, it organises and orchestrates the market participants on the supply and demand side and provides the infrastructure for a successful development of the ecosystem to providers and consumers.

9.6.3 *Start Small to Go Big*

From the customer's point of view, based on the collected data from the platform, a significant increase in efficiency in the areas of software selection and implementation can be assumed, since the necessary connections for the implementation of digitalised asset management are already available at platform level and can be provided with little effort. The establishment of a SaaS marketplace for digital wealth management also gives providers a high degree of flexibility regarding the design of their business model and the necessary infrastructure. It will be much more cost-effective to test certain innovations first in a smaller, defined target group, as the start-up costs for digital business models will be a fraction of today's usual project costs.

9.6.4 *Once a Competitor, Now an Ecosystem Partner*

From the platform operator's point of view, the strategic orientation is lucrative if it is possible to unite a relevant number of partners and customers on the marketplace and to facilitate simplified access to software components from different providers with the help of professional business process management. The participation in such a SaaS market makes sense for software providers in case the reduced resource requirements for the acquisition of new customers are invested in further development of one's own offering, strengthening the competitive position with respect to other providers with similar functional scope. Hence, making the provider attractive for the customers regarding price and/or functional scope. In order to establish such an ecosystem, significant investments by the marketplace operator and the SaaS providers are required to create the necessary technical and procedural prerequisites. The effort seems reasonable if the ability to innovate, speed and efficiency of innovation can be materially increased.

CHAPTER 10

How Can Robo-Advisory be Implemented and Integrated into Existing Banks?

Ana-Maria Climescu, Christian von Keitz, Jan Rocholl, and Madeleine Sander

Contents

A.-M. Climescu (✉) • C. von Keitz • J. Rocholl • M. Sander
Hauck & Aufhaeuser, Frankfurt am Main, Germany
e-mail: ana-maria.climescu@hauck-aufhaeuser.com; christian.keitz@hauck-aufhaeuser.com; jan.rocholl@hauck-aufhaeuser.com; madeleine.sander@hauck-aufhaeuser.com

© The Author(s) 2021

P. Scholz (ed.), *Robo-Advisory*, Palgrave Studies in Financial Services Technology, https://doi.org/10.1007/978-3-030-40818-3_10

10.1 Background

The following case study deals with the question of how robo-advisory can be implemented and integrated into the existing bank model, using the example of Hauck & Aufhaeuser Privatbankiers (H&A). In this context, we understand robo-advisory as an online application that offers automated financial advice and services.

Firstly, we will look into the bank's background. As part of its strategy, H&A has set itself the goal of examining the extent to which the use of new technologies can improve existing business and unlock new market potential across all divisions. This gave rise to the following question with respect to private banking, which is the focus of this chapter: "How can we use the new technical possibilities to address changing customer needs as well as generate new business opportunities and revenue streams?" To answer this, we first analyzed client needs before turning our attention to the investment market.

10.1.1 Introduction

H&A was created in 1998 as a result of a merger between two well-established institutions: the Frankfurt-based Bankhaus Georg Hauck & Sohn (founded in 1796) and the Munich-based Bankinstitut H. Aufhaeuser (founded in 1870) (Hauck & Aufhaeuser 2020c). H&A is thus one of the oldest private banks in Germany with a history stretching over 220 years.[1] In addition to serving private clients, the company has specialized over the past three decades in financial services for professional investors and businesses. H&A divides its activities into four divisions: private banking, asset servicing, financial markets and investment banking (Hauck & Aufhaeuser 2020a).

The bank currently has 722 employees[2] across a total of seven locations in Germany, Luxembourg and Switzerland.[3] In addition to its headquarters in Frankfurt, H&A also has offices in Munich, Hamburg, Duesseldorf and

[1] In June 2017, H&A was transformed from a KGaA (partnership limited by shares) into the legal form of an Aktiengesellschaft (joint-stock company); see Hauck & Aufhaeuser (2017a, p. 34).

[2] Annual Report 2019, Hauck & Aufhaeuser (2019a).

[3] In addition, H&A has a subsidiary in London and Nanjiing as well as a representative office in Paris.

Cologne. While its German branches mainly focus on private banking, the Zurich branch specializes exclusively in ethical and sustainable investment aimed at private and institutional investors. Nevertheless, Luxembourg has recently become the bank's largest location. This is due, on the one hand, to the expansion of both the depositary and fund management business and, on the other hand, to the acquisition of the Luxembourg-based Sal. Oppenheim entities in 2017/2018 (Hauck & Aufhaeuser (2016) and Hauck & Aufhaeuser (2017b)).

This step fits seamlessly into its corporate strategy, "H&A Strategy 2020", which focuses on growth and digitalization. This was adopted in 2017 by the Executive Board, the Supervisory Board and the new Chinese owner Fosun following the takeover (see Fig. 10.1). The takeover by the Chinese group is one of the most important milestones in the bank's recent history. In mid-2015, the bank officially announced that the Chinese investor group Fosun plans to acquire a majority stake in the bank. The official transfer of shares took place in the fall of 2016.

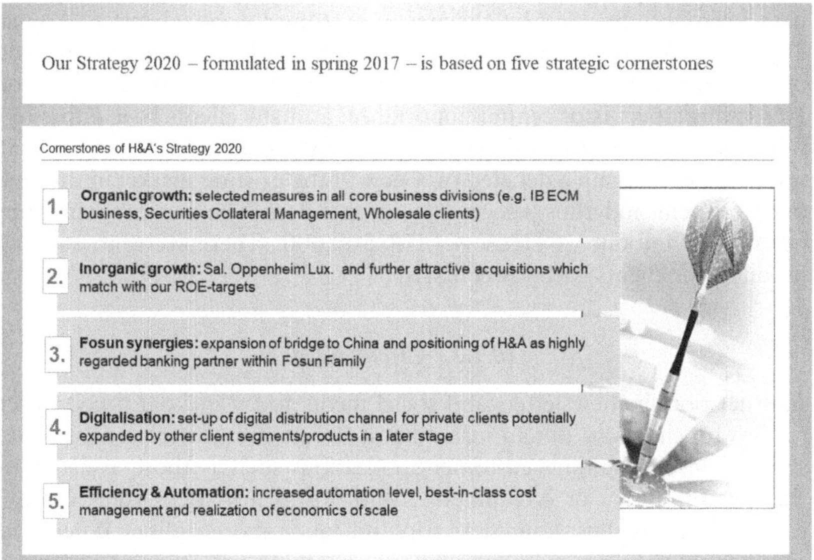

Fig. 10.1 H&A strategy 2020 (taken from presentation slide)

H&A is the first German private bank to be majority owned by a Chinese investor (Mussler 2016). In addition to the analysis and evaluation of organic and inorganic growth potential, the strategy is to tap into the cross-selling potential between China and Germany. However, the focus of the digitization project is on expanding the range of services offered to customers and increasing customer benefit through intuitive applications. This also involves modernizing the bank's internal infrastructure and automating administrative work processes to spend more time advising customers. At the same time, concrete measures have been defined for each business division that contribute to the overarching corporate objective of developing H&A into a state-of-the-art, market-leading bank in the German-speaking region.

In the first step, the "H&A Strategy 2020" focused on the evaluation of strategic, organic growth options for private banking. Initially the demand of private clients for banking services was analyzed from various perspectives. For private banking, which is regarded as "people's business", the behaviour and wishes of clients play a particularly important role. Digitalization has opened up new technological possibilities, some of which have been accompanied by changes in customer needs (Ernst & Young 2019, pp. 65). Many aspects of the world in which we live and the needs of the next generation have been through a period of profound change. For H&A, this aspect is of central importance, as many clients have entrusted their assets to the bank for the third or fourth generation. In addition, it was necessary to gain a detailed overview of the existing market in the area of investment and thus also of the competition. The current minimum investment amount of H&A is EUR 500,000. There were no suitable products for clients with lower assets.

10.1.2 Client Needs

Internet portals, newsletters and social media today make it much easier for private clients to obtain information. In addition, there are intuitive, simple and entertaining customer applications such as those available in the travel industry or e-commerce. With just a few clicks you can now order books, clothes, your next trip and much more online. When you are shopping for a suit online, you get matching belt recommendations and when you are browsing for an investment guidebook, you can gain very interesting insights into the buying behavior of others, for example, through the "Other customers also bought" window. According to trend research (Linden and Wittmer 2018, pp. 1, 7, 13, 16), the gigatrend of digitalization and the resulting megatrends such as individualization, knowledge culture and mobility will further increase the demand for such services.

In-house client feedback shows that this is also reflected in the expectations of private customers when it comes to financial investments—starting with the desire to have an overview of the performance of their assets at all times in real time, through to the possibility of being able to compare or directly conclude individualized proposals during the investment process, regardless of time or place, and thus avoid going to the branch during normal business hours.

At the same time, it is interesting to note that studies show that existing private banking clients, especially those with growing wealth, do not consider the level of digitalization or, surprisingly, the price to be among the top criteria determining their satisfaction. Instead, they value the service received from the client advisor, along with asset preservation (Nicolaisen 2018). As a result, it can be noted across the board that personal contact and proximity to the advisor as well as performance and product variety remain relevant across all customer segments (Deloitte 2015, pp. 24–25). In addition, the brand's reputation is particularly important for traditional private banking clients (l.c.). While user experience and convenience are already among the top factors determining corporate customer satisfaction, existing private banking clients tend to be more reserved in this respect (l.c.). At the same time, however, the increasing use of technology in everyday life and the associated stronger affinity for digital applications obviously mean that having fast and convenient access to a bank account is becoming even more important to private banking customers (l.c.). Somewhat surprisingly, reputation and exclusivity of the brand are seen to be gaining ground in this context (l.c.). This suggests that positioning in the private banking market will in future range between:

1. convenience and technology
2. service quality
3. brand reputation

The demand for technology and innovation and thus the importance of the first factor is increasing among clients, whether it is today's average client or the inherited wealth client, and is even more pronounced among the under-40 age group (Nicolaisen 2018). This is where fintechs or robo-advisors come in. Having its origins in America with pioneers like Betterment and Wealthfront, the robo-advisory market is growing fast in China and gaining in importance in Germany, even though the market forecasts vary somewhat and have to be viewed critically based on the actual growth rates to date: Statista (2019d) expects the investment volume in Germany to reach EUR 24 billion by 2022, representing an annual growth rate of 48% (compound annual growth rate (CAGR) 2019–2021). Roland Berger, a global consulting firm, anticipates a somewhat more conservative compound annual growth rate of 20–40% compared to 4–6% for conventional asset and wealth management (Buess et al. 2018, p. 8). The comparison with the conventional segment clearly underscores the immense potential of this new market segment.

But what exactly is behind this potential? How should H&A respond to this development? Offering ease of use (e.g. when opening an account), a high degree of transparency and, on balance, attractive conditions, robo-advisors firmly have their finger on the pulse of changing customer behavior as described above. Specialist studies have also come to the conclusion that "hardly any of the established asset managers will be able to do without robo-advice in the future". At the same time, the average investment volume, depending on the provider, is around EUR 10,000 (fintego)[4] or around EUR 32,000 (Scalable Capital 2018), which is well below the volume generated by H&A's current target client segment. The question here is, therefore: to what extent is this development relevant to H&A and its target client segment?

These considerations pose major challenges to the banking sector in general and the private banks such as H&A in particular. As a medium-sized wealth management boutique steeped in tradition, the bank has often demonstrated its innovative strength and resistance to crises. Unlike large corporations, however, private banks do not have multi-billion dollar budgets available to fund their development. Although open to new ideas, many bankers were also skeptical about the robo-advisory market at first. Personal communication and client engagement are key success factors in sales at H&A. H&A builds long-standing relationships with clients, knows their families, their goals and interests, and not only looks after

[4] Based on ebase report (European Bank for Financial Services GmbH 2019, p. 6).

their investments but also acts as a "sparring partner" for topics such as career planning, raising children, succession planning or new trends, and often has an interest in similar hobbies. Acquiring new customers whom you have never even spoken to before sounds a bit strange, especially in a private bank.

In addition to these strategic and cultural issues, the growth in infrastructure of established private banks over the years presents another challenge. Without a fundamental transformation, they cannot compete against the "greenfield" fintechs and their plug-and-play[5] approach based on application programming interfaces (API). Both old core banking systems and decentralized data repositories are among the biggest obstacles when it comes to speed. In addition, new methods such as agile project management or the approval of minimum viable products (MVP), which are already commonplace in the fintech and robo-advisory sectors, are still seen as uncharted territory by established banks.

10.1.3 Market Analysis

In addition to customer needs, market potential also played a key role in H&A's private banking strategy. Specifically, the question was whether the use of technology in the existing client segment above EUR 500,000 could lead to greater customer satisfaction and thus attract a greater number of new customers, or whether a significant growth potential could be identified in the other client segments. In order to validate these scenarios, the first step was to have an in-depth overview of the investment market (i.e. market volume and medium-term market forecasts) and the existing providers (or the range of products and services offered by existing robo-advisory market players) to explore the strategic options for H&A and, based on this, develop a clear position for the bank's own approach.

The market potential and volume are determined primarily by the financial assets of private households. In Germany—the primary target market for H&A's private banking activities—household financial assets amounted to around EUR 6053 billion at the end of the third quarter of 2018 and rose by around EUR 76 billion or 1.3% compared to the second quarter of 2018 (Statista 2019c). Taking a five-year view, the growth

[5] The English term *plug-and-play* refers to the possibility of connecting new devices to a computer without having to install additional drivers or changing the settings.

Private Banking / Wealth Mgmt. segments	Liquid assets	Number of households	Share of the market for private customers	Compound annual growth rate 2018 - 2022
Wealth Management clients	≥ 3,000 TEUR	~ 63,500	0.16 %	4 – 6 %
Private Banking clients	500 to < 3,000 TEUR	~ 560,000	1.4 %	1 – 3 %
Affluent Banking clients	300 to < 500 TEUR	~ 737,000	1.8 %	1 – 3 %
Retail Banking clients	< 300 TEUR	~ 39,300,000	96.0 %	-1 – 1 %

Fig. 10.2 Private banking client segments in Germany 2018 (based on zeb.research)

rate was around 15%, on the basis of, EUR 5222.6 billion in the third quarter of 2015 (l.c.). In 2018 (Statista 2019a), German investors favored conventional savings (39%), followed by fund units (34%), equities (26%), real estate (26%) and call money (26%). The savings rate, which has been rising steadily since the last financial crisis, also appears interesting in this context (Statista 2019b). At this point, it is important to break down the financial assets of private households even further and to take a closer look at the relevant market segments in relation to their market potential (Fig. 10.2).

A private banking study by zeb expects the CAGR or compound annual growth rate of the private banking market in Germany to be 4%. In more detail, wealth management clients are forecast to grow by 4–6%, closely followed by private banking clients with 1–3%. The forecasts for affluent and retail clients are more restrained: 1–3% for affluent banking clients and −1 to 1% for retail banking clients. In this context, however, it is important to consider the market share of the respective client groups. While the market shares of wealth management clients, private banking clients and affluent banking clients are 0.16, 1.4, and 1.4%, respectively, retail banking clients account for the bulk of the market share with 96%. H&A currently serves private banking clients and wealth management clients. Here it is necessary to consider how technology can be used to improve or, where appropriate, expand the existing range. It is therefore necessary to take a closer look at the affluent and retail segments to assess any untapped potential for the bank.[6]

Another study by BCG reveals significant weaknesses in the affluent segment, in particular, with regard to the range of services currently offered. At the same time, the study ascribes the greatest potential for future growth to this segment: For example, investable assets in the global affluent client segment—defined here as those with assets ranging from USD 250,000 to USD 1 million and comprising 76 million people—are expected to grow at an above-average CAGR of 6.2% over the next five years (Zakrzewski et al. 2019, pp. 12–13). This segment has a great market potential. Although retail banks, private banks, discount brokers, insurance companies and fintechs have developed a broad range of products, the affluent segment continues to be underserved (Zakrzewski et al. 2019, l.c.). It is dominated by expensive products from well-known brands or simplified offerings from retail banks that do not meet the needs of affluent clients (Zakrzewski et al. 2019, l.c.). In order to successfully serve this client segment and develop the associated market potential as a bank, it is necessary to offer products that meet the specific needs and preferences of affluent sub-segments (Zakrzewski et al. 2019, l.c.). In specific terms, this means expanding services and products to include the right technology. A bank must be seen as a hybrid business model that combines digital and human engagement to provide clients with a more personalized, convenient and at the same time successful asset solution and performance (Zakrzewski et al. 2019, l.c.).

BCG has closely examined the weaknesses of existing providers in the affluent segment (Zakrzewski et al. 2019, l.c.). Although competition for this segment is increasing, most banks and asset managers have not yet found an answer to their clients' needs. **Private banks** with a focus on wealth management tend to overserve their affluent clients at the beginning of the relationship, offering them their own relationship manager and processes designed primarily for wealth management clients. But when the relationship fails to live up to its perceived potential, the bank may be tempted to reduce its engagement. From a cost perspective, this is of course understandable, but may leave clients feeling neglected. Private banks with a focus on affluent clients have in some cases barely done anything to tailor

[6] Data for this paragraph taken from Nicolaisen (2018).

their product range (e.g. broader product range) and service offering (e.g. digital access) to client needs and are thus hardly in a position to offer sophisticated financial or investment advice to a wide range of affluent clients.

Retail banks are still more focused on the sales and reach of their brands and products than on the specific needs of this client group. Although **fintechs** or robo-advisors have the advantage over their rivals in innovation, they do not have the market access, reputation or the trusted advisor status enjoyed by private banks. **Insurance companies** are also trying to tap into this promising segment. However, they lack the resources and the necessary expertise to develop individual asset-management solutions. Some **asset managers** have again tried to expand their range of services by buying and deploying robo-advisors, with mediocre results so far. That said, some providers have regrouped after their first efforts and have begun to work on digital additions to existing investment models that should ultimately allow advisors to serve this segment better and more cost-effectively. **Brokerage firms** have largely underserved this segment from a product perspective, focusing too much on short-term trading and too little on long-term wealth creation. Also here, there are some signs of improvement.

Overall, however, this market remains poorly served and thus offers great earnings and market potential to an established traditional brand such as H&A. The last step was therefore to take a closer look at the current market for digital offers in the field of asset management and the client groups addressed. It is also noticeable that even established private banks, but also new robo-advisors, are focusing on the retail rather than the affluent segment. The positionings of current market participants tend to follow two paths: either they emphasize the level of digitization, which in the case of robo-advisory is usually tied to a smart algorithm, or they emphasize the proximity to the bank and the brand behind it. This also depends on whether the companies are fintechs or robo-advisors (e.g. Scalable, ginmon, or Liquid) and thus mostly relatively young companies or a bank spin-off or the digital asset management of an already established bank (e.g. Solidvest, Warburg Navigator or Castell Insight). There are also offshoots of banks that, like fintechs, consciously position themselves through automated risk profiling in conjunction with semi-automated rebalancing.[7] The competitors relevant to H&A can thus be grouped into three clusters (Fig. 10.3), which, from a positioning perspective, range between "smart algorithm" and "proximity to a parent company" (see Fig. 10.3). For H&A's private banking, the question was how to leverage its strengths in positioning its digital offering and how it can differentiate itself from other providers. Following in-depth studies of client needs and an analysis of the market environment, it has become clear that there is a great potential for H&A in the affluent segment, which has so far received only rudimentary service from other providers. As this is a group of affluent clients who have been denied access to the bank's products and services because of their assets, the reduction in the minimum investment requirement for the bank is a logical consequence of the new technological possibilities. H&A not only has the necessary expertise in the field of investment but also the necessary reputation to harness the potential of the affluent segment.

10.2 The Implementation and Integration Process

The following key findings can be formulated based on the market and customer analysis:

- The following three features are crucial for positioning in the private banking market:
    1. convenience and technology
    2. service quality
    3. brand reputation
- The demand for technology and innovation is increasing among clients, whether it is today's average client or the inherited wealth client, and is even more pronounced among the under-40 age group.
- This is where robo-advisors come in, which are expected to grow faster than traditional private banking providers because they offer a better user experience, greater transparency and more attractive terms.

[7] Rebalancing is defined as the regular adjustment of a portfolio in line with the defined investment strategy.

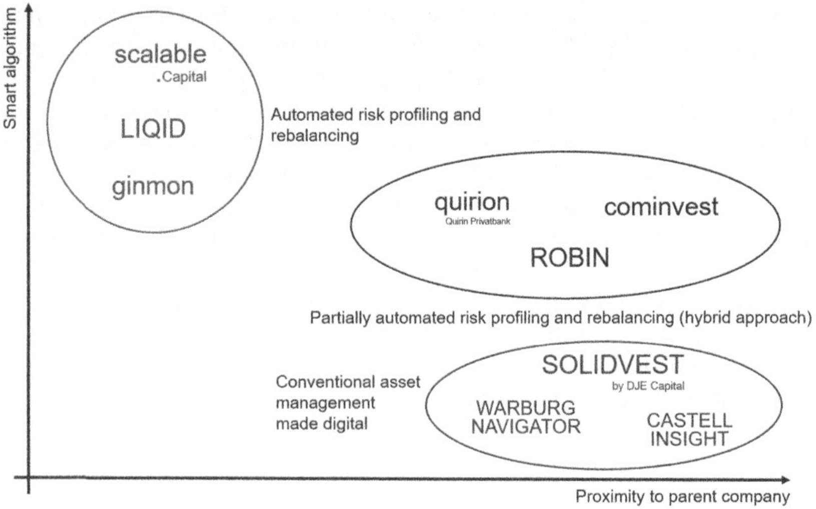

Fig. 10.3 Products and services offered by existing market providers in the robo-advisory segment

- The analysis of the client segments shows that, although the private banking market is expected to grow faster, the percentage share of the affluent and retail segments is much bigger.
- While most financial investment providers concentrate on the retail segment, the affluent segment is underserved by established financial service providers and robo-advisors.
- Existing robo-advisors have so far focused on their positioning on the level of digitization or the proximity to the parent company/parent brand.

These findings then flow into the development of growth scenarios and possible next steps for H&A's private banking, to answer the overarching question at the heart of the case study. Robo-advisory services are introduced on five[8] levels: strategy, product, marketing and sales, regulatory and the technical platform.

[8] See Accenture (2019).

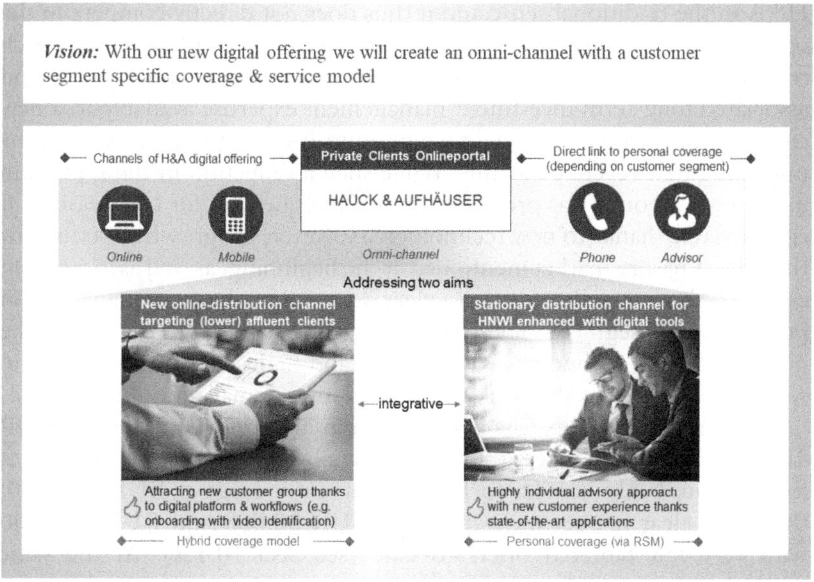

Fig. 10.4 Strategic objective of an integrated omni-channel

10.2.1 Strategy

The omni-channel[9] emerged as a result of the key findings of the market and client analysis, long-standing expertise in working with private clients and the analysis of strengths of H&A in a seamless transition between a new digital sales channel to be established for affluent clients with a minimum investment of EUR 50,000 and a local sales network for conventional private banking customers with a minimum investment of EUR 500,000, through personal advice offered in the branches and supported by modern applications. H&A's strategic approach is explained in more detail below (Fig. 10.4).

The market and client analysis showed that the robo-advisory business model offered to date by fintechs is less tailored to the needs of private banking and wealth management clients serviced by private banks such as

[9] In this context, omni-channel means a cross-channel business model, e.g. the combination of high-street and digital sales.

H&A in the traditional sense and it thus does not directly compete in the core group segment. At the same time, the technology-based approach represents a smart way of making H&A's asset-management solution and associated long-term investment management expertise available to a new client group below the previous minimum investment amount, thereby unlocking new revenue streams. While such a reduction in the minimum investment amount was previously out of the question for cost reasons, it is now viable thanks to new technology. However, the growth forecasts for the robo-advisory market mentioned at the beginning, as well as local client contacts, show that there is also a clear demand for the asset-management product in this segment. There was particularly strong interest among the new generation, technology-savvy clients and high-net-worth clients who would like to try out the bank first by investing a small amount.

While most robo-advisors with minimum investment amounts of EUR 10,000 or even less concentrate on the retail segment and the offering of local private banks, similar to H&A, starts on average at EUR 500,000, there is a clear market niche between EUR 50,000 and EUR 500,000 with very few targeted offers to date (see Sect. 10.1.3). At the same time, the asset situation in the affluent segment is associated with other needs and investment opportunities. In this area, it is already possible to invest in individual securities (in contrast to pure ETFs or funds) without jeopardizing adequate diversification. As an asset manager for high-net-worth clients, H&A has been well-aware of this need.

A comparison of the company's strengths with the market potential and the competitive situation revealed that H&A's new digital sales channel should target the affluent segment, as the segment with the greatest potential. In addition, internal analyses have shown that this segment plays a key role for H&A when it comes to attracting the next generation of private banking and wealth management clients. H&A also realized that its brand and the expertise of its client advisors count as particular strengths, potentially giving the bank a competitive edge when acquiring new clients in this segment (see Sect. 10.1.2). This resulted in the omni-channel strategy (see details below) with a digital distribution channel supplemented by a hybrid advisory approach and a seamless transition to the local distribution network with increasing complexity of financial circumstances.

Even though the analyses showed that the pure robo-advisory business model had a lesser effect on traditional private banking or wealth management client segments, in-house client surveys carried out at the same time have shown that the technological demands of clients have risen considerably in recent years as a result of digitization (with respect to online banking, among other things).

The first phase of the project therefore focused on modernizing the set-up of the local sales channel and the needs of the existing client group—for example, by implementing a new, modern online banking system with real-time access to asset portfolio (including detailed analysis of asset structure, e.g. by asset class, country or industry). This included measures such as the introduction of a WhatsApp news channel and a virtual tour of the branch offices on the bank's website. The measures were based on client and market analysis results as well as a client survey carried out in advance in local sales. The project prioritization reflected the need for a modern online banking system for the digital sales channel and the central requirement of addressing the needs of existing client groups. In line with the methods of agile project management, H&A relied on MVP for the introduction of the new online banking, that is, a product that meets the minimum requirements of its customers. In contrast to conventional development methods, this approach allowed clients to see the first technological improvements after just a few months. This basic version was then equipped with additional features in subsequent releases, and it is being continuously developed. Thanks to the agile approach, it was possible to obtain feedback from clients based on MVP and to incorporate client feedback with respect to ease of use and new requirements into the prioritization for the next project phases.

The second step consisted in developing a completely new concept for digital asset management, Zeedin.[10] This included all aspects of a market launch—starting from the product and service range, including pricing conditions, putting together a team responsible for this and creating the interface with existing private banking and other departments, developing the underlying operational processes and the technical platform through to targeted marketing and communications.

The following three strategic questions played a central role for H&A in determining the range of services:

- What would an intelligent interaction between a human and a machine look like?

[10] The reasons for naming the service "Zeedin" are discussed in detail in the Sect. 10.2.3.

- What options does H&A have to ensure that its product range meets the increasing demand for personalization?
- How can H&A provide its customers with personal access in the digital channel despite the cost pressures?

As a result, the bank opted for a hybrid approach, that is, an intelligent interaction between human beings and technology. Before we get into further details, it is important to understand the basic functions and features of Zeedin: the starting point is Zeedin's website, which offers prospective clients a comprehensive overview of the services Zeedin offers. This is followed by an introduction to the range of investments offered by Zeedin and information about the related investment process, which is divided into three steps:

- In the first step, the prospective client is required to complete an online questionnaire to determine their personal investment profile, taking into account their investment objectives, risk appetite and existing financial knowledge.
- On this basis, Zeedin selects an optimal investment proposal for the prospective client in the second step. Where required, the investor can further personalize the investment proposal, for example, by deselecting individual asset classes or choosing the investment universe. In line with the high level of information in local sales, Zeedin's digital investment proposal is also very detailed and includes, in addition to the strategic asset class allocation, a simulation of future performance, historical performance information, regional and sector allocation and top portfolio picks.
- In the third step, the client can choose between a fully digital process using the latest video authentication technology and a semi-digital variant with authentication in a post office. In both cases, the necessary personal data is collected and the contract documents based on this data are made available through Zeedin's digital application process.

Zeedin's product range stands out thanks to the following exclusive features, state-of-the-art technology and expertise: through Zeedin, clients who have EUR 50,000 or more to invest can now access H&A's multi-

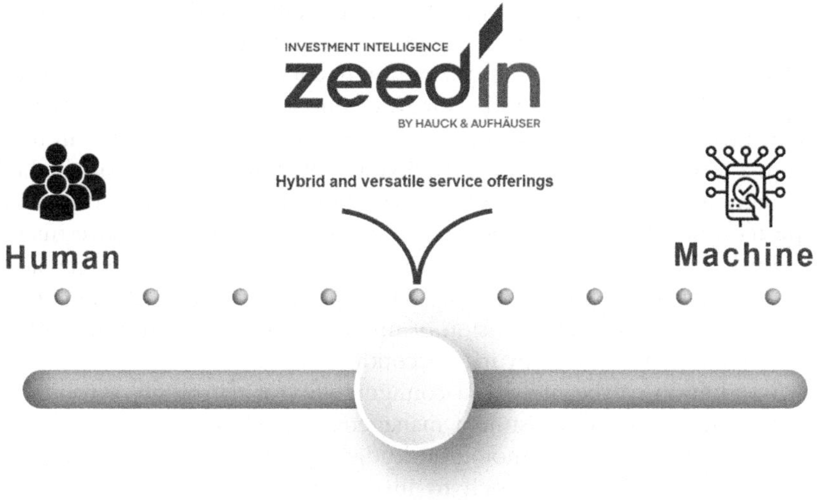

Fig. 10.5 Intelligent interaction between people and technology

award-winning[11] asset management product, developed over many years, which until recently was only available to high-net-worth clients with EUR 500,000 or more to invest. In line with H&A's mission statement of "Achieving a return through active management across all liquid asset classes" (Hauck & Aufhaeuser 2020b), Zeedin's investment management approach also involves a combination of strategic asset class allocation specified by the client online and a tactical asset allocation depending on the market phase. The tactical asset allocation and the selection of the securities in which to invest is then carried out by H&A's investment management experts based on a structured process. H&A also relies on Zeedin's combination of human and computer intelligence with respect to the ongoing monitoring and adjustment of the portfolio (Fig. 10.5). In this way, any portfolio adjustments are transparent and can be retraced by the client at any time, and can also be explained based on fundamental valuations carried out by our investment managers.

[11] Received the Golden Bull—"Asset Management of the Year 2019" award from the magazines €uro, €uro am Sonntag and Boerse Online, and the transparency in asset management award—"R&P VV Ausweis" from Roedl & Partner.

In addition, Zeedin stands out through its ability to adapt to the various requirements of the target client group. This begins with the product selection: Zeedin offers clients the option to choose initially between three types of discretionary portfolio management solutions (see Sect. 10.2.2) depending on the required investment universe. Overall, clients can pick from a total of over 100 individual investment strategies based on further configuration options (such as the deselection of individual asset classes or specification of the equity allocation) in conjunction with the investment and risk profile derived from the questionnaire. The service range can also be customized through optional additional services such as a portfolio review in the first year, provision of monthly macroeconomic analysis or participation in six-monthly macroeconomic webinars.

In addition to the individual configurability of the range of products and services and the investment management track record supported by human expertise and technological leadership, another key advantage of Zeedin is its hybrid approach to investment advice. Based on offline sales experience[12] and the findings from Sect. 10.1.2, it was clear that personal contact should not be underestimated with respect to digital sales channels. Zeedin's clients and prospective clients have access to a dedicated customer service team—via email, chat or telephone. In addition, prospective Zeedin clients have the opportunity to discuss their investment approach with a client advisor from local private banking before signing a contract. Here H&A is again relying on modern applications such as a digital appointment calendar for arranging meetings and co-browsing for joint, location-independent discussion of the investment approach.

With the resulting positioning and the omni-channel approach involving Zeedin and local sales, the bank is pursuing several strategic goals:

- a hybrid and transparent approach—a prospective client can, for example, obtain comprehensive information via the website and draw up an individual investment proposal using the Zeedin application route without obligation.
- a customer-specific approach with a range of products and services tailored to the individual needs and financial circumstances of the client,

[12] Based on interviews with client advisors and analysis of the conventional customer journey in private banking at H&A.

- a seamless transition between the two sales channels, including an integrated overview of the company's website, the design of the service range and the platform (both channels use the same online banking).

In principle, the following also applies: the higher the investment volume, the more individual and personal the range of services. The range of services offered by H&A's traditional private banking—from an investment volume of EUR 500,000—includes additional key elements such as inheritance planning, further personalization of asset management (e.g. through the option of deselecting individual industries or considering regular distributions) and investment advice as an alternative to discretionary portfolio management. In addition, clients with an investment volume of EUR 500,000 or more are assigned a permanent personal advisor.

10.2.2 Product

As part of the implementation of the digital asset management platform Zeedin, it was important to reconcile several goals from the strategic perspective. The main focus, however, was on unlocking the potential of a new client group in the affluent segment, designed to complement H&A's focus on the private banking segment.

Furthermore, it should be based on the proven investment approach of the portfolio management of the bank, that is, the bank would seek to develop a complementary digital sales channel rather than an algorithm-driven selection process. The product range should be transparent in relation to all costs incurred and provide clients with convenient access from any location. As H&A wants to achieve a positive alpha factor compared to the benchmark for its clients through targeted stock picking, Zeedin should also offer a variant with individual stock selection. In addition, the bank can look back at a long tradition of ethical and sustainable investment: in 1995 H&A launched the first and thus the oldest ethical fund in continental Europe. This expertise should also be reflected in Zeedin's product range (Schraner 2011). In order to offer a financial investment without the relatively high minimum investment amounts required for the individual security variants from as little as EUR 50,000, the offering was expanded to include an ETF-based solution (Fig. 10.6). These unit-linked asset management and individual security solutions should allow clients to deselect mandatory components such as gold and risk-adjusted investments (certificates with a negative correlation effect as a portfolio supplement). As already mentioned, transparent and regulation-compliant pricing was an important aspect (under MiFID II, in particular, banks are required to provide a much greater level of detail from 2018 and illustrate the resulting reduction in returns). From the client's perspective, all relevant factors must be clearly visible and comprehensible. If possible, this should not give rise to any additional costs. For this reason, a pricing model has been developed for the client which includes all third-party expenses, administrative costs, transaction and depositary costs of the bank as well as any currency conversion expenses. VAT at the applicable rate is also taken into account in accordance with the requirements of the Price Indication Ordinance. In addition, some financial instruments used incur product-related costs (e.g. running costs related to funds and certificates). These tend to be specified by nearly all market participants, but the level of detail of this information varies considerably among providers. In Zeedin's case, for example, the bank not only specifies the running costs and the depositary bank fee for funds (these are usually listed in the prospectus as a total expense ratio, but do not take into account the transaction costs incurred within a fund). Even though transaction costs for ETFs are typically low, active funds can still add substantial positions that are not always disclosed to the investor.[13] These are therefore calculated and disclosed under costs in EUR and as percentage of the investment amount in accordance with the requirements of MiFID II.

10.2.3 Marketing and Sales

Once the product had been defined at the strategic level and the development started at the operational level, the bank had to plan the customer journey[14] and thus the marketing concept, in order to position Zeedin's

[13] Certificates may incur product-related costs when purchased through the issuer, where the fair value of the product differs from the subscription price. We have also selected and specified these costs.

[14] In this context, the customer journey refers to the entire customer experience with the product—from the first contact through marketing through examination of the product to conclusion of the contract and subsequent use of the online banking platform.

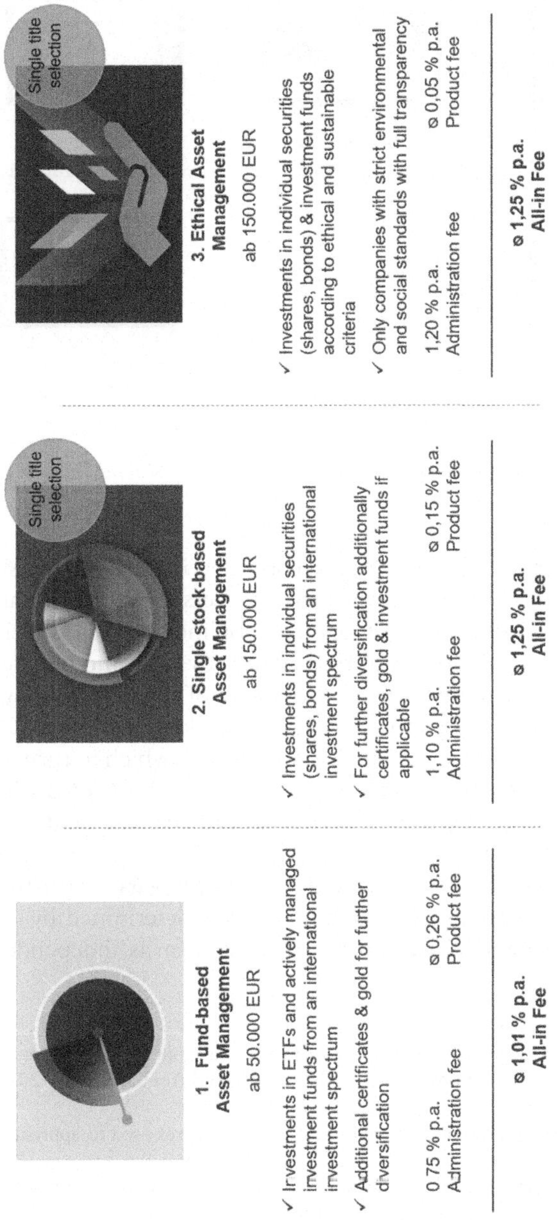

Fig. 10.6 Zeedin product range (H&A webpage)

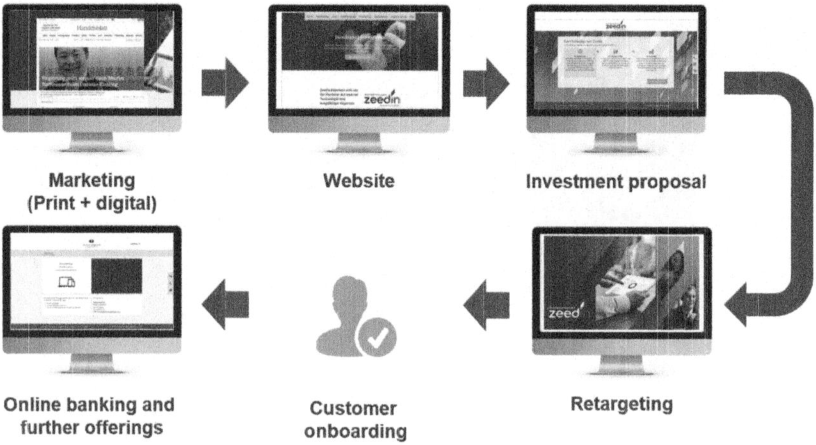

Fig. 10.7 Customer journey with Zeedin

USP on the market.[15] The customer journey is divided into six process steps: marketing, website, investment process/proposal, retargeting,[16] onboarding/securities account opening, online banking and other services. The online banking platform had already been modernized.[17] This sub-project also included the development of the onboarding process. Marketing for Zeedin therefore focused on marketing the product in order to lead potential customers to their own website, which in turn leads them to the investment section and finally to the creation of an investment proposal. The next step was to define a lead management process that would enable systematic retargeting (Fig. 10.7).

The development of the marketing concept began with the detailed analysis of target groups. Although it is largely determined by the volume of initial investment, characteristics that are just as independent of the

[15] The entire marketing strategy for Zeedin was developed in collaboration with the Munich-based communications agency Saint Elmo's; see Annual Report 2018 Hauck & Aufhaeuser (2018).

[16] In this context, retargeting describes marketing measures used to approach a potential customer again.

[17] Zeedin's onboarding process is technically linked to H&A's online banking to provide customers with a consistent product experience.

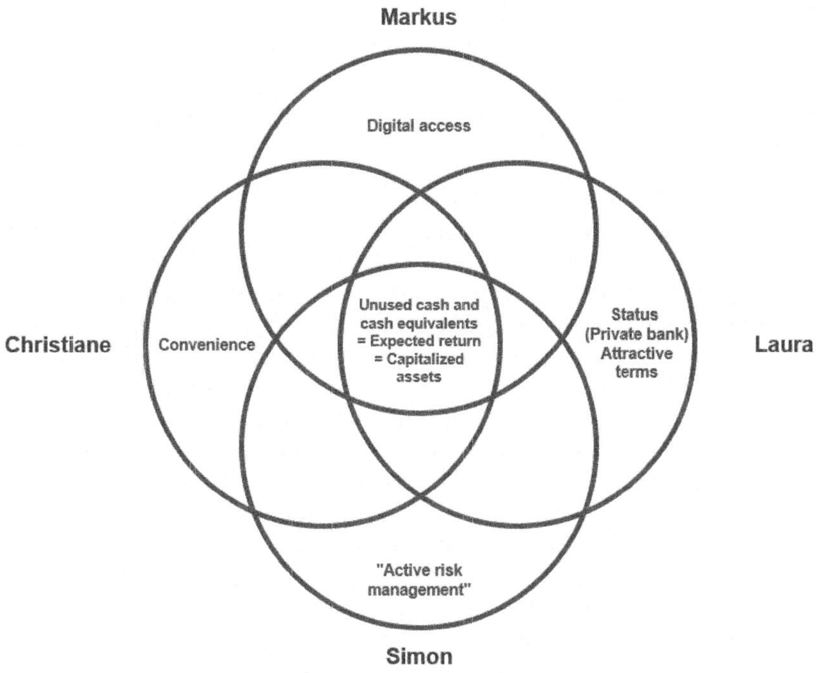

Fig. 10.8 Analysis of customer needs

assets such as occupation, education or hobbies, play a role when it comes to a targeted and individual approach. In the preparation of the so-called personae, the bank used empirical values and studies of the needs of potential clients. This process has generated four personae, whose individual lifestyles were outlined: Markus, Laura, Simon and Christiane.[18] The personae were then superimposed with one another. Although the trigger varies from person to person (digital access, convenience, "private bank" status, attractive terms and "active risk management"), a common feature quickly became apparent: the needs of the target group were based on large amounts of unused cash (Hauck & Aufhaeuser 2019a) (Fig. 10.8).

[18] These names were chosen at random and represent the personae used internally.

Fig. 10.9 Zeedin logo

The overlap of the target group consists of the fact that the assets remain unused for a longer period of time and are actually reduced by negative interest rates and inflation. H&A's new product should offer an innovative solution to activate the untapped potential of cash assets to invest based on individual ideas. This insight should then be reflected in the entire history of the brand. Firstly, the bank needed to find a suitable name. At a strategic level, the name had to reflect the bank's multifaceted approach and the combination of different values and perspectives. Therefore, the team has to ensure that the name did not sound too technical, but also human, and thus reflected the connection of the best of both worlds as the core of a hybrid identity. A number of expectations were thus attached to the new product name Zeedin (Fig. 10.9).

The word Zeedin [siːd ɪn] is intentionally based on the English word "seed in/seeding" and should refer to the planting/growing/sowing aspect. The term *seed* or *zeed* stands for growth and profit, future, spirit of the time and flexibility, while *in* stands for intelligence, innovation and individuality. On this basis, the logo was developed, and the bank added the subtitle "Investment Intelligence by Hauck & Aufhäuser" to reflect the client demand for a well-known, exclusive brand, proximity to the bank and the bank's own private banking business. This is also reflected in Zeedin's corporate design (CD), for example, in the investment process (Fig. 10.10).

Based on the fundamental elements of the CD, it was necessary to develop a campaign that would not only draw the attention of the target clients to the product on launch, but also beyond that, and at the same time pick them up at their point of pain, which is the unused cash. The messages "Money is not stupid" and "If you invest intelligently" were developed for this purpose. Each personae (Markus, Laura, Simon and Christiane) was assigned an individual motif that transfers the message of the campaign into their individual environment (Fig. 10.11).

A further focus of Zeedin's marketing concept was the design and implementation of the website (Hauck & Aufhaeuser 2020d). In contrast

Fig. 10.10 Layout of Zeedin's investment process

Fig. 10.11 Visualizations—personae and messages

to other H&A services, this is a sales site that not only enhances the image of the bank but also aims to motivate potential clients to go through the investment section and prepare an investment proposal. At this point, the bank had to decide whether to create a separate website or to integrate it into the H&A corporate website. For this purpose, all competitors with a close proximity to a banking house were examined.[19] It was noticeable that none of them had an integrated website, although all providers advertised their new digital offerings on their respective corporate websites. This was likely due to organizational reasons, like technical hosting, or strategic reasons.

For Zeedin, an integrated approach was chosen after weighing all the advantages (e.g. positive impact of the parent brand H&A on digital asset management and thus greater trust on the part of clients, potential linking of marketing activities, integration into the existing website is associated with lower technical costs) and disadvantages (e.g. proximity to the parent brand must always be maintained; the marketing presence must not counteract other messages, reputation risk if digital asset management does ends in failure), not least in order to highlight the proximity to offline private banking and the hybrid advisory approach. Today, when website visitors click on private banking, they can choose between conventional and digital asset management. The website itself is divided into a total of six sections: Home, Investment strategy, Expertise, Products & Conditions, Security and Frequently asked questions (FAQ). An additional "Login" button allows returning visitors to re-enter the investment process by entering their login details at the point where they left when they last visited the website.[20,21] The implementation of the Zeedin section on the H&A website also focused on transparency and information. The costs and terms & conditions are always just a few clicks away. The individual subpages

[19] This refers to banks which, like H&A, have expanded their offline presence by adding a digital sales channel.

[20] Before the prospective client can view their investment proposal, they are required by law to enter their personal data. Next, the prospective client receives a password from the investment section, which they can use to login again at any time and change their data.

[21] As soon as the prospective client enters their personal data into the Zeedin investment section, they become part of the lead management process established and developed specifically for this purpose. This then allows a personalized approach to the prospective client/client (depending on the status) at any point in the customer lifecycle.

have their own FAQ area, which provides answers to the most important questions visitors may have (Hauck & Aufhaeuser 2020d).

In addition to the website as the central sales platform, the bank also had to develop an adequate communications mix for Zeedin. The creation of high-quality content, that is, content marketing, was the first priority. This involved the creation of various articles about strategy and the functions and features of Zeedin, together with a high-quality video.[22] Due to the digital affinity of the target clients, the marketing was focused at the beginning on targeted online marketing measures: Google, social media, Amazon, display ads and cross-channel retargeting. These measures were complemented by a print campaign and earned media (e.g. interviews). The advantages of the integrated marketing approach quickly became apparent: The launch of the product on October 26, 2018, was spread across all H&A's communication channels, enabling Zeedin to benefit from the bank's existing followers.

A further focal point for publicity was Zeedin's presence at trade fairs and selected events. The aim here was to find out which trend themes were of interest to the different personae, for example, "women and finance". In addition to the conventional marketing approaches, the corporate network of the investor Fosun was also involved from the outset.[23] For example, Zeedin's first customers were offered a discount on a Club Med trip (Aufhaeuser 2018). Later, this cooperation was also extended to Club Med customers. Depending on their status, club members of the travel group have the opportunity to benefit from various discounts when concluding an asset management contract with Zeedin.[24] The integration of Zeedin into the ecosystem of the investor of H&A Fosun[25] thus increases synergies on both sides, which ultimately benefit the clients of all the companies involved.

10.2.4 Regulatory Requirements

For the project team at H&A, it was important from the outset to consider all regulatory requirements and where applicable, ensure that they are

[22] Available at (Hauck & Aufhaeuser 2020d).

[23] Fosun has also invested in Club Med since 2015: (BBC 2015).

[24] See Hauck & Aufhaeuser (2019b).

[25] Since 2016 H&A has been majority owned by the international investment group Fosun.

fully complied with. As a financial services institution with a full banking license, H&A is subject to supervision by the German Federal Financial Supervisory Authority (BaFin). In this context, the bank is required to undergo a comprehensive audit yearly conducted by an external and independent auditing firm. The results of this audit are presented in a consolidated report and made available to the supervisory authority. The concept therefore took into account the very broad scope of this audit in order to meet all documentation requirements.

A digital service not only has to pay attention to national regulatory requirements but also to the regulatory requirements in the countries of potential clients. When selling products or services to persons residing in the USA, for example, the bank would have to comply with the far-reaching US consumer protection law, which in some cases deviates significantly from European legal standards. Therefore, in the first step, the bank had to narrow down the potential target client group. To keep the complexity of the implementation in the first phase (initial product development) manageable, the bank decided to focus initially only on clients residing in Germany. To avoid complex tax structures in the initial phase, the bank also decided to open only individual client portfolios for the time being. The bank planned from the outset to make this service available to clients residing in the European Union in the next phase.

The bank is, in particular, required to comply with MiFID II and the General Data Protection Regulation (GDPR) of the European Union.[26] While the provisions of the MiFID II Directive are not directly binding, as they are transposed into the respective national laws, GDPR as an EU regulation is directly binding. In early 2018, the far-reaching changes to MiFID II came into effect, and they are thus relevant to the project. Compared to other EU countries, Germany has already anticipated a number of national requirements in previous years, so that we can speak here of an even stronger, established, regulated market.

While the analysis in this case study focuses on the investment recommendation aspect of a given implementation form, it disregards the far-reaching requirements for actual portfolio management (as there have not been many innovations that would be particularly worth mentioning compared to non-digital distribution channels). The requirements for the

[26] Regulation (EU) 2016/679 of the European Parliament and of the council of 27 April 2016.

client acceptance process or further topics such as the prevention of money laundering will not be discussed here.

While the implicit meaning of the term robo-advisor is automated advice, the definitions still vary. However, from a legal perspective, this does not constitute investment advice, as this investment service under the German Banking Act (KWG) explicitly and exclusively targets the recommendation of a particular financial instrument, such as a security or a fund (Bundesanstalt für Finanzdienstleistungsaufsicht & Deutsche Bundesbank 2019, p. 1). However, the recommendation for a financial service such as portfolio management—as in this case—is not subsumed under the KWG. Although this does not constitute investment advice under the relevant rules and regulations, the MiFID II requirements set out in the Securities Trading Act (WpHG) stipulate that financial institutions must nevertheless ensure that the recommendation is suitable for the client. This means that knowledge and experience in securities transactions must be available in a relevant form or must be conveyed as part of the advice/recommendation and the investment must be suitable for the client, taking into account the client's financial circumstances, investment objectives and individual risk tolerance (see Article 64 (3) WpHG). As already mentioned, the bank is regularly checked for compliance with these requirements as part of the so-called WpHG audit (this strict annual audit does not exist in other European countries or only to a limited extent),[27] so the bank is required to produce the relevant documentation.

According to initial hints from BaFin, it can be assumed that the recommendations from robo-advisors could in the future fall within the scope of investment advice (Bundesanstalt für Finanzdienstleistungsaufsicht 2017). From the client's perspective, robo-advisors also provide a service comparable to investment advice by reviewing very personal information and offering specific recommendations. At the same time, clients do not distinguish between the various regulatory requirements.

The Zeedin project team has therefore decided to take a proactive step and to apply the same high regulatory standards to this new digital sales channel and the recommendation of a specifically designed asset management solution as the bank would apply to personal discussions with discerning private clients at any of the bank's branches . For this reason,

[27] Exceptions include e.g. Austria, where financial institutions must submit to regular independent audits.

the client questionnaire and investment objectives are also recorded in the relevant suitability report in accordance with MiFID II, as would be required when providing investment advice to private clients with respect to individual financial instruments.

The complete digitalization and automation of this process was only made possible by the introduction of MiFID II. By the end of 2017, financial institutions in Germany had to prepare an advisory protocol for investment advice. This document was intended to provide the client with a tool to reflect the investment advisory situation and the main arguments before closing the transaction and, in case of doubt, enable reconstruction of the conversation afterwards. This process involved recording of every individual information about the conversation. In addition, precise formal requirements did not allow for a purely technical recording. For example, the advisory protocol always had to be signed by hand by the relevant investment advisor.

With the implementation of MiFID II, this instrument has been removed from the German WpHG and replaced by the suitability report. This has much lower formal requirements (e.g. no signature) and rather than being a protocol of a conversation, it gives substantive reasons for a recommendation. As a result, it is now also possible to reproduce this document completely by technical means.

These considerations had a direct influence on the necessary design of the digital sales channel. In order to be able to ensure that the suitability report can also be submitted on time, it is mandatory to provide an e-mail address before the recommendation is displayed to the client. This means that the document can always be verifiably transmitted as soon as a recommendation has been made, even if the potential client simply aborts the process or closes the browser. Since Zeedin offers a very individual product range, it was also necessary to ensure that any changes to client information resulted in a new, updated suitability report. Clients can, for example, log into an existing consultation again and change their details if necessary, which would result in a modified recommendation. In this context, it was necessary to define which personal data should be collected during customer evaluation. The data obtained in this way can then also be used to address clients correctly and to transfer them to the onboarding process. Information such as education level and field of activity of the client, which is collected and taken into account for investment advice, was designed in such a way that this information can also constitute mandatory information when the client makes an investment.

All the GDPR requirements had to be taken into account, which meant that we had to limit ourselves in the aforementioned information to what is absolutely necessary to make a recommendation. The relevant data erasure/retention periods were also implemented and the information about our data protection measures was adjusted accordingly to reflect information about the new sales channel.

10.2.5 Technical Platform

The development of Zeedin and the online banking backend infrastructure could not be started with a greenfield approach, but had to be based on existing and central banking applications. Zeedin's front-end application, on the other hand, could be developed and designed independently and according to the latest insights to enhance the customer experience. The technical challenge here was the seamless integration of state-of-the-art technologies and specialized banking service providers or fintechs into the established banking systems and processes, in particular the core banking system.[28]

A key challenge here was to enable external access to the real-time information of the core banking system (master data, account and securities transactions). The previously closed core banking system, which supplied the original online banking system exclusively with data from batch-based overnight processing, had to be opened up for web access via a secure infrastructure. This was the only way to meet the requirement for an automated and direct new client onboarding process when opening a Zeedin securities account, that is, direct exchange during interaction with a new customer.

The core banking system, which was a legacy system from the technological standpoint, was extended to include a modern web services layer, in this case a REST API. As a result, all relevant business transactions, from new customer investments to online banking processes such as foreign transfers, were available to third-party systems in real time. In addition, this architectural approach allowed sensitive data to be kept strictly within the bank while integrating robotic and online functionalities from external service providers.

[28] Client accounts and securities accounts are managed in the core banking system, among other things.

The integration of web services into the core banking system was new territory for the bank. In addition to the technical security requirements for setting up the network and firewall infrastructure, the technical aspects of implementing a sophisticated, multidimensional authorization concept for online clients also had to be taken into account.

The latest security standards were implemented as part of the infrastructure development, the effectiveness of which had to be demonstrated by various penetration tests. Communication between the servers was secured through the integration of server certificates. In addition, users authenticate themselves with client certificates, which also enable encrypted data transmission (using TSL).[29] As is customary in the industry, the transactions of bank customers are protected by a two-factor authentication system, which is being further expanded under PSD2.

To develop the digital onboarding with direct entry of all client master data including accounts, securities accounts and online access to the core banking system, numerous processes had to be automated as required by law. The KYC process, which is usually run asynchronously and semi-automatically in conventional and stationary processes, is particularly worthy of mention here. For the first time, the KYC check had to be carried out without any disruption with minimal latency as a regulatory prerequisite for opening an account. To do this, the relevant compliance application was connected via REST API and following video legitimation, the checking process was automatically triggered and its results evaluated. Here too, real-time access via web services was implemented for the first in the relevant application using real-time API. As a result, not only has the technology been upgraded but the business process flow of the KYC check, which is based on it, has also been fundamentally changed.

For the first time, the chosen technical architecture allows online customers to access H&A's document management system via SOAP web services.[30] Current and historical client records such as account statements can be made available in an electronic mailbox at any time without having to create memory-intensive copies of the documents for online access and

[29] TSL stands for transport layer security, and it is a protocol designed to provide communications security over computer networks.

[30] SOAP stands for "Simple Object Access Protocol", which can be used to exchange data between systems or to call a remote procedure.

redundant data storage. The documents remain stored at their source at all times in a manner that ensures data integrity.

In addition, both the Zeedin application and online banking from H&A offer modern technologies for communication between end customers and the bank. An external partner provides both a cloud-based co-browsing solution[31] and live chat functionality, which are seamlessly integrated into the website. As a result, the client communication is "web-enabled", which means that investment products requiring explanation can be discussed with the client on a split screen.

Through the consistent implementation of real-time interfaces and the elimination of batch processes,[32] it is possible to complete client onboarding processes without any disruption or manual interactions. This process integrates industry-leading specialist vendors with features such as KYC[33] compliance checks or videoident procedures. For initial asset-allocation purposes, the bank integrated a service that calculates the minimum variance portfolio depending on the risk class and based on the Markowitz (1952) method.

The complex infrastructural development and redesign of Zeedin's business processes as well as the upstream introduction of modern online banking as a digital hub show how to move successfully into a new, strategic business area through an optimal combination of customer focus, design thinking,[34] agile implementation methods and the targeted use of smart technologies.

10.3 The Real Question Is: "How Can the Bank Use the Latest Technology to Expand Its Range of Services in the Interest of the Client?"

Many aspects had to be taken into account when answering this question, resulting in a multifaceted approach. Although the question of

[31] With co-browsing, two or more people can navigate the same website at the same time and can mark certain places or use visual aids such as arrows to show them to the client.

[32] These are process chains that are queued together to form a *batch*.

[33] KYC stands for "Know your customer", which is a criminal background check in accordance with Article 8 of the 3rd EU Anti-Money Laundering Directive.

[34] Design thinking is a method for developing projects and innovations.

the appropriate technical implementation (especially the "make or buy" decision[35]) and the integration into an existing bank played a key role and the implementation required a dedicated project team, the task went far beyond the introduction of a technical application. Rather, the question of strategic positioning and the needs of the target client group ran as a thread through the entire project.

During implementation, it was also necessary to parallelize task packages and coordinate numerous internal and external experts from IT, marketing, sales, compliance, and so on. For instance, the application process was completed in several two-week sprints[36] developed in parallel to the account opening process and the website. In this context, it was important not to lose sight of the integrated customer journey (see Sect. 10.2.3; Fig. 10.7) to avoid any process-related breaking points between different applications. This demanded high levels of communication and teamwork as well as quick and straightforward decision-making processes. The entrepreneurial spirit often associated with small and medium-sized businesses, combined with a high level of personal engagement, a sense of responsibility and a certain willingness to take risks—that is, a team committing to an idea as if it was their business—played a central role in the implementation process. However, this only works if the team is fully trusted by the management and given the appropriate decision-making authority.

It was also important to see things consistently from the client's perspective: Which pain point drives our target client group? Where does the client expect the latest technology to be deployed to help him manage his finances? In which areas does the client consider personal interaction to be essential? Which factors can strengthen trust, in particular, on the digital customer journey? These and other questions took up a lot of space alongside the regulatory requirements and technical feasibility. With the omni-channel strategy and the digital asset management platform Zeedin, H&A has opted for an approach that gives customers a choice in key areas. Based on the results of client and market analysis (see Sect. 10.1.2 Client needs and 10.1.3 Market analysis), the bank has given priority to those parts of the customer journey where clients require information (e.g. information about the range of services and terms offered by H&A

[35] This refers to the decision as to whether a company prefers to produce something itself (make) or to buy something from a third party (buy).

[36] A sprint is one of several cycles within an agile project.

provided on a transparent website, non-binding and advisor-independent preparation of an investment proposal using Zeedin, daily, location- and time-independent monitoring of the development of the portfolio using modern online banking across all end devices). Zeedin offers clients the option to discuss the investment strategy with a client before signing a contract, which is the key decision point in the investment process. Accordingly, the experienced investment team from H&A is responsible for picking the securities for the portfolio and its day-to-day management, assisted by state-of-the-art portfolio management systems rather than automated algorithms. Zeedin clients can thus benefit from the existing track record of investment experts managing the funds of high-net-worth clients. Daily client interactions show that historical performance of real client portfolios is a central element in building trust and thus also a key factor in the decision-making process.

After market launch, it quickly became clear that any strategy is only as good as the team that lives by it every day. There were many soft factors and personal incentives that played a decisive role in the consistent implementation of the omni-channel approach. In addition, close personal interaction between the teams was essential for success.

References

Accenture. 2019. *Omnikanal Vertrieb im Banking*. Last Accessed 24 March 2019. https://bankingblog.accenture.com/deutsch/omnikanal-vertrieb-im-banking?lang=de_DE.

BBC. 2015. *China's Fosun Wins Bidding for Club Med After Two Years*. Last Accessed 20 March 2020. https://www.bbc.com/news/business-31432322.

Buess, Robert, Daniel B. Hildebrand, Dominik Löber, and Jeanette Okwu. 2018. Rewiring Asset and Wealth Management, P. 8. Munich: Roland Berger GMBH.

Bundesanstalt für Finanzdienstleistungsaufsicht. 2017. *Robo-Advice – Automatisierte Anlageberatung in der Aufsichtspraxis*. BaFin Journal. 18–22. https://www.bafin.de/SharedDocs/Veroeffentlichungen/DE/Fachartikel/2017/fa_bj_1708_RoboAdvice.html.

Bundesanstalt für Finanzdienstleistungsaufsicht, and Deutsche Bundesbank. 2019. *Gemeinsames Informationsblatt der Bundesanstalt für Finanzdienstleistungsaufsicht und der Deutschen Bundesbank zum Tatbestand der Anlageberatung*. Last Accessed 20 March 2020. http://www.bafin.de/SharedDocs/Downloads/DE/Merkblatt/dl_mb_110513_anlageberatung_neu.pdf.

Deloitte. 2015. *Wealth Management and Private Banking – Connecting with Clients and Reinventing the Proposition Value*. https://www2.deloitte.com/

content/dam/Deloitte/ch/Documents/financial-services/ch-en-financial-services-efma-wealth-management-private-banking-032015.pdf.

Ernst & Young. 2019. EY Banken-barometer 2019–Zeichen der Zeit. Tech. rep. 2019. https://www.ey.com/Publication/vwLUAssets/ey-bankenbarometer-2019-broschuere/\protect\LY1\textdollarFILE/ey-bankenbarometer-2019-broschuere.pdf.

European Bank for Financial Services GmbH. 2019. *Robo Advice in Deutschland – Deutschland – Status quo und Entwicklungsperspektiven*. ebase. P. 6.

Hauck & Aufhaeuser. 2016. *Hauck & Aufhauser Acquires Luxembourg-Based Sal. Oppenheim Asset Servicing Business from Deutsche Bank*. Last Accessed 20 March 2020. https://www.hauck-aufhaeuser.com/en/hauck-aufhaeuser-acquires-luxembourg-based-sal-oppenheim-asset-servicing-business-from-deutsche-bank.

Hauck & Aufhaeuser. 2017. *Annual Report 2017*. Tech. rep. Last Accessed 24 March 2019. https://www.hauck-aufhaeuser.com/fileadmin/Publikationen/Geschaeftsbericht/180606_HA-Jahresbericht-2017_DE.pdf.

Hauck & Aufhaeuser. 2018. *Mit Zeedin 20% Rabatt auf Ihre Club-Med-Reise*. Last Accessed 20 March 2020. https://www.hauckaufhaeuser.com/mit-zeedin-20-rabatt-auf-ihre-club-med-reise.

Hauck & Aufhaeuser. 2019. *Annual Report 2019*. Tech. rep. Last Accessed 1 August 2020. https://www.hauck-aufhaeuser.com/fileadmin/Publikationen/Geschaeftsbericht/GB_ENG_BANK.pdf.

Hauck & Aufhaeuser. 2019. *Club Med*. Last Accessed 24 March 24 2019. https://www.hauck-aufhaeuser.com/ad/club-med.

Hauck & Aufhaeuser. 2020. *About Us*. Last Accessed 20 March 2020. https://www.hauck-aufhaeuser.com/en/about-us.

Hauck & Aufhaeuser. 2020. *Anlagekonzept und Investmentansatz*. Last Accessed 20 March 2020. https://www.hauck.aufhaeuser.com/zeedin/anlagekonzept.

Hauck & Aufhaeuser. 2020. *Hauck & Aufhaeuser–History*. Last Accessed 20 March 2020. https://www.hauck-aufhaeuser.com/en/about-us/company-profile/history.

Hauck & Aufhaeuser. 2017. *Hauck & Aufhauser Successfully Completes Acquisition of the Luxembourg-based Sal. Oppenheim Companies*. Last Accessed 20 March 2020. https://www.hauck-aufhaeuser.com/en/hauck-aufhaeuser-successfully-completes-acquisition-of-the-luxembourg-based-sal-oppenheim-companies.

Hauck & Aufhaeuser. 2018. *Produktangebot im Bereich Private Banking um eine Digitale Vermögensverwaltung*. Last Accessed 20 March 2020. https://www.hauck.aufhaeuser.com/hauck-aufhaeuser-erweitert-mit-zeedin-sein-produktangebot-im-bereich-private-banking-um-eine-digitale-vermoegensverwaltung.

Hauck & Aufhaeuser. 2020. *Zeedin kümmert sich um Ihr Portfolio.* Last Accessed 20 March 2020. https://www.hauck-aufhaeuser.com/zeedin.

Linden, Erik, and Andreas Wittmer. 2018. *Zukunft Mobilität: Gigatrend Digitalisierung und Megatrends der Mobilität.* Universität St. Gallen. P. 1, 7, 13, 16. https://www.alexandria.unisg.ch/253291/2/Zukunft%20Mobilit%C3%A4t%20%20Gigatrend%20Digitalisierung_A5_final.pdf.

Markowitz, Harry M. 1952. "Portfolio Selection". *Journal of Finance* 7(1): 77–91.

Mussler, Hanno. 2016. *Chinesen Dürfen das Bankhaus Hauck & Aufhäuser Übernehmen.* Last Accessed 20 March 2020. https://www.faz.net/aktuell/wirtschaft/unternehmen/fosun-darf-das-bankhaus-hauck-aufhaeuser-uebernehmen-14383183.html.

Nicolaisen, Christian. 2018. *Margen im Private Banking Sinken auf Kritisches Niveau.* Last Accessed 20 March 2020. https://www.private-banking-magazin.de/zeb-studie-2018-margen-im-private-banking-sinken-auf-kritisches-niveau-1.

Scalable Capital. 2018. *Scalable Capital Durchbricht Milliardengrenze.* Last Accessed 20 March 2020. https://de.scalable.capital/presse/scalable-capital-durchbricht-milliardengrenze.

Schraner, Remy. 2011. *Es Fehlt eine Echte Bereinigung.* Last Accessed 20 March 2020. https://www.n-tv.de/deleteMe/migration/Es-fehlt-eine-echte-Bereinigung-article4475046.html.

Statista. 2019. *Aktuell genutzte Geldanlagemöglichkeiten.* Last Accessed 24 March 2019. https://de.statista.com/statistik/daten/studie/838/umfrage/aktuell-genutzte-geldanlagemoeglichkeiten/.

Statista. 2019. *Entwicklung der Sparquote privater Haushalte seit 1991.* Last Accessed 24 March 2019. https://de.statista.com/statistik/daten/studie/2699/umfrage/entwicklung-der-sparquote-privater-haushalte-seit-1991.

Statista. 2019. *Geldvermögen der Privathaushalte.* Statista. Last Accessed 24 March, 2019. https://de.statista.com/statistik/daten/studie/37880/umfrage/geldvermoegen-der-privathaushalte-in-deutschland/.

Statista. 2019. *Robo-Advisors: Deutschland.* Last Accessed 20 March 2020. https://de.statista.com/outlook/337/137/robo-advisors/deutschland.

Zakrzewski, Anna, Tjun Tang, Galina Appell, Renaud Fages, Andrew Hardie, Nicole Hildebrandt, Michael Kahlich, Martin Mende, Federico Muxi, and Andre Xavier. 2019. *Global Wealth 2019 Reigniting Radical Growth.* Tech. rep. Last Accessed 30 June 2019. Boston Consulting Group. http://image-src.bcg.com/Images/BCG-Reigniting-Radical-Growth-June-2019_tcm9-222638.pdf.

PART IV

The Future of Robo-Advisory

CHAPTER 11

The Role of Artificial Intelligence in Robo-Advisory

Alexander D. Beck

Contents

A. D. Beck (✉)
Quoniam Asset Management, Frankfurt am Main, Germany
e-mail: alexander.beck@quoniam.com

© The Author(s) 2021

P. Scholz (ed.), *Robo-Advisory*, Palgrave Studies in Financial Services Technology, https://doi.org/10.1007/978-3-030-40818-3_11

11.1 Introduction to Artificial Intelligence

Owing to the highly digital and scalable nature of web-based asset management, artificial intelligence (AI) and robo-advisory have a lot of potential to go hand in hand in manifold ways. However, even though AI is a strong and powerful technology, it is not by default the right choice for any emerging business challenge. As fields of application need to be suitable in certain aspects, they need to be deliberately selected. The following chapter describes areas in robo-advisory where AI can be used with good chances of success, but it also points to impediments and limitations.

From the perspective of information processing, AI describes a technology that automatically detects patterns in all sorts of data. These patterns are correlated to the likelihood of certain (future) events to occur. Based on these likelihoods, the AI system will make decisions and take actions, such as suggesting a product and selling a stock. The list of examples where AI is utilized today is long and diverse: It ranges from the detection of icebergs from satellite images to the prediction of the next best movie to watch. Among the most well-known applications of AI are Amazon's product recommendation service or Apple's Siri. Hidden from the perception of retail customers, AI is used for optimizing logistics processes or marketing, sales, and service events. Regarding robo-advisory, the playing ground for AI appears quite large, ranging from customer-related topics to portfolio management. Predicting the stock market with AI is an exciting but challenging field of application: Historic price information and other market and non-market data are utilized to predict future price developments (Domingos 2012; Hastie et al. 2009).

From a bird's-eye perspective, AI emerged from the field of statistical and machine learning. It adds some human-like behavior to "pure" machine learning (which usually takes place on structured, tabular data only). As such, AI is often referred to when humanoid cognitive functions are imitated by a computer, such as machine vision and machine-based text processing. Examples include estimating the meaning of a written text or determining its sentiment.

With respect to the machine learning part of AI, many different modeling approaches are utilized and continuously further explored. Two of the most prominent are artificial neural networks and random forests, just to name a few (Breiman 2001).[1] These modeling techniques are powerful methods for describing real world non-linear and complex phenomena, and they usually outperform classical modeling techniques such as linear regression in terms of prediction power. Non-linear phenomena need to be well-considered in modeling because they are impactful and all around us. Black swan events and panic reactions at the financial markets are prominent examples.

From a practitioner's perspective, AI falls into two categories—general (or strong) AI and applied (or weak) AI. General AI denotes a technology which can perform any intellectual task the human brain is capable of. This type of AI still lies in the future and is today not yet existent. On the other hand, applied AI denotes a technology that is used to perform specific tasks. One example of applied AI is the Statoil iceberg challenge (see kaggle.com), where an AI algorithm is utilized to distinguish icebergs from ships based on satellite imagery. As it turns out, what is not possible for the human eye is feasible for a well-trained AI algorithm. As such, applied AI tends to outperform humans when it comes to such highly specialized problems. In the course of this chapter, only applied AI is considered.

11.1.1 Historical Development of AI in the Investment Management Industry

The philosophical roots of AI date back hundreds of years and even philosophers like Aristotle had early ideas of "mechanical thinking" (Giles 2016). A conference at Dartmouth College in 1956 laid the foundation of today's state of AI. AI was a much-hyped field of research in its early years, and machines on a human intelligence level were expected within a few tens of years. Obviously, these estimates drastically underestimated the real academic effort and necessary computational power to achieve this goal. AI research suffered from two funding crises (also known as "AI—Winters") in the late 20th century due to unmet expectations in the development progress. Today, AI has reached a productive state where its application is a natural part of many software projects. However, a human-like intelligent machine is still more science fiction than reality.

[1] Cf. also Hinton, who won the Turing Award (Association for Computing Machinery 2018).

When it comes to financial markets, machine learning, a subfield of AI, and especially neural networks have quite a history in the field of predicting the financial markets. Even though it appears very tempting and promising to throw financial market data at some strong AI model, the reality of predicting financial markets looks somewhat different: Financial markets are dominated by randomness to a large extent, and it is notoriously difficult to squeeze out the predictable part of market movements. After some hype around applying expert systems (a subfield of AI) to the financial markets applications, practitioners were rather disappointed by the performance of this approach. The reasons behind this frustration could be either of the following: (1) highly complicated systems or (2) the unmet expectation that AI would render human intellectual input into the system needless (Durkin 2002). However, two lessons from these days are still valid today: First, "crap in crap out", i.e. (financial) data needs to be at its best quality and deliberately selected and second, overfitting is always just one tiny step away (overfitting happens when a model is so flexible that it adjusts to random fluctuations on the training data and cannot discover the actual correlations any more). However, in retrospect, today's even stronger models would have significantly outperformed markets in the past (Fischer and Krauss 2018). This might lead to the assumption that financial markets are constantly balanced at the level of what forecasting technology allows at each time point.

Driven by the latest technological advancements in data availability and model performance, AI models are again in vogue in the financial industry.

11.1.2 What Can be Expected from AI Today?

Today, the field of AI has strongly evolved, and additional aspects of modeling markets as well as predicting customer behavior come into play. On the one hand, the increase in available computational power at a comparatively low cost allows the development and utilization of more sophisticated AI models. On the other hand, data is collected more extensively than ever before and is now widely available, many times even for free.

Moreover, the application perspective has changed: During the past decade, the sophisticated scientific component of utilizing AI increasingly disappeared behind easy-to-use technology frameworks. Today, it is even possible to assemble AI systems by simply dragging together modules on a graphical user interface. Due to this development, AI has become easily accessible to a wide range of users, comparable to a car driver who does not need to have working knowledge of the car engine in order to drive. Naturally, this encourages a broader use of AI especially in areas where returns on investment need to be achieved after short development times. In the highly digital field of robo-advisory, this leads to a fruitful playing field for AI. Especially in digital business models and in situations where decisions are made repeatedly in a high frequency, the game-changing character of AI will become increasingly visible.

11.2 Value Potential for AI Along the Core Processes of Robo-Advisory

In a strongly digital business model such as robo-advisory, process automation and decision optimization are obvious levers behind business success. Especially, customer-related processes that repeat themselves frequently are perfectly suited for applying AI-driven decision support. Such processes are found along the customer journey in manifold situations. Due to the digital nature of robo-advisory, each customer journey creates an electronic footprint. This footprint is very valuable from a business analytics perspective, and its proper recording and data management is the foundation of a data-driven and AI-optimized customer experience.

11.2.1 Customer Data and Collection Strategies

Customer data form the foundation for applying AI to improve customer processes. They are collected across touchpoints of the customer journey, and typical customer data sources include the following:

- Master data (annual income, gender, age, pension plans, assets like real estate, family situation, etc.)
- Marketing interaction data (interaction with past marketing campaigns, reaction to outbound calls, participation in surveys)
- Website/app interaction data (after login, which parts of the website or the app are especially important to the customer? Where do customers spend a lot of time?)
- Service data (inbound calls, service requests, type of service requests)
- Product purchases (transactions)

For companies with data spread across different (legacy) systems, it is an expensive challenge to map the above-mentioned data sources properly to a single customer. However, this investment pays out once customer-related data is used to improve the customer journey. With respect to the technical aspect of the data collection process, it is important to choose only tools and services that allow easy access to the stored customer information on a single-user granularity. Additionally, data quality is of high importance, and the right quality-assurance measures and futureproof data designs should be in place as early as possible in the robo-advisor's life. Another important aspect of customer data is data privacy, which should be in line with regulation.

Due to its digital and scalable nature, a robo-advisor is not limited to a maximum number of customers, and AI-driven customer actions play a crucial role for the robo-advisor's success. Obviously, when the number of AI use cases grows across the marketing, sales, and service landscape, and the customer base grows as well, increasing computing power is needed. The technological foundation should not be a limiting factor on this growth path. Consequently, it is important to build all data and analytics systems as scalable designs and on futureproof infrastructure (typically cloud infrastructure).

Generating a data-driven customer view requires an appropriate data-collection strategy throughout the whole customer journey. For some business models, this is a serious challenge: For example, a typical problem of insurance companies is that they do not hear much from their customers after closing a contract until certain payout events occur. This is a poor data basis for managing customer processes by AI. Robo-advisory service providers, on the other hand, have much better possibilities of interacting with their customers on a regular basis. Easy accessibility, for example, by providing a smartphone app and offering interesting ways of interacting with the portfolio management service, creates points of attraction and makes the service sticky for its users. This paves the way for receiving important information from clients on a frequent basis. For example, it is a key task for a robo-advisor to gauge its clients' risk appetite regularly. This figure will change over time, and this information enables the robo-advisory service to adjust individual portfolios accordingly. Thus, the customer is always provided with an appropriate investment plan and has a feeling of safety.

To make the customer journey even more attractive, the approach of gamification can be used to induce enjoyable and playful interaction elements. This increases the willingness of customers to use the service and spend some time on things like customer surveys (Seaborn and Fels 2015). For example, clients' reactions to political events might be used to gauge their risk sensitivity and awareness in a playful way.

11.2.2 Marketing, Sales, and Service

Marketing, sales, and service are typically the three pillars of customer relationship management. All three have great potential to be drastically improved by applying automated decision-making with the help of AI. In the digital world, and from the robo-advisor's perspective, these three pillars benefit from each other and should be strongly intertwined. Considering the client's perspective, interconnections lead to a smooth and enjoyable customer journey. Imagine, for example, a service request that leads to an offer for a better product experience without reaching out to the sales department separately. From a technological viewpoint, the interconnection between the three pillars takes place naturally. Think, for example, of a product recommendation engine which not only suggests products and generates sales but also offers service touchpoints by suggesting what previous customers found helpful when purchasing a certain investment product.

Especially in marketing, AI has a huge potential for increasing the return on marketing expenditure. Here is an example: Assume a cross-selling campaign where an incentive is paid to those customers who add a risk management module to their robo-advisory service. How can AI help in this scenario? At first sight, this problem appears to be a plain and simple propensity-scoring problem; however, it is more complicated than that. Four types of customers need to be distinguished:

1. Customers who would have added the risk-management module either way because they are risk-averse and seek ways to minimize risk.
2. Customers who are being convinced by the campaign and purchase the risk management module only because of the incentive.
3. Customers who receive the campaign and see it as a reminder to cancel their portfolio.
4. Customers who will not enter the deal no matter what.

The modeling approach to tackle this challenge is called uplift modeling. Only those customers who are triggered to buy through the incentive are selected by the model and all others are left untouched. Typically, returns on investment in uplift-modeling campaigns turn out to be significantly larger than with other approaches, because the target is so precisely defined and in line with monetary goals. Such an uplift model can be constructed with classifying neural networks or other statistical learning models. Most important is the data-generation part, on which such a model is constructed. The data needs to be generated on a pilot campaign carried out on a randomly drawn subset of the customer base. Only by assuring randomness in the customer-selection process, a bias-free training data set can be achieved.

The role of AI in this scenario is to correlate the available customer data to the uplift. One advantage of this method is that customer data needs to be explicitly collected in pilot studies for this approach, which means there is no need to depend on existing customer data. A sufficiently large number of customers (typically > 1000) needs to be available for this approach.

11.2.2.1 Chatbots

Chatbots are another example of how the customer journey can be supported with the help of AI. Chatbots have emerged over the past years (e.g. Microsoft's Chatbot Service "LUIS" was launched in late 2017 (Hernandez 2017)) and provide an interesting way to utilize written speech recognition to support customer interactions with AI.

Human-based customer service is typically quite expensive and is difficult to scale. Web-based services like robo-advisory that target a digitally affine group of people should consider utilizing digital service channels and employing their human service reps for qualified cases only, that is, where an answer can only be found or prepared by a human. Consequently, chatbots offer a way to scale customer services quite effectively. Chatbots provide various ways to support customer interactions (Cava 2016):

1. Directly communicating with a customer and answering straightforward questions. Example: "Where can I find my YTD portfolio performance?"
2. Providing links for self-help in case of less well-defined questions and handing over to a service rep if the answer is unsatisfactory for the client. Example: "How can I find out which portfolio is right for me?" -> "Website: Portfolio Finder"

3. Preparing a conversation with a service or sales rep: Example: "Please provide your name and email address, and you will be directly handed over to the next free assistant."
4. Being supportive to a service representative. While the service representative talks to a client, she can use a bot service simultaneously, for example, to quickly find reference cases. Example: "Tax issues for a German robo-advisor when the customer is a US tax-payer."
5. Guiding through interactions with multiple levels, for example, guiding through a process of signing up to a service. Example: Purchasing an investment product where certain investor criteria need to be fulfilled.

Chatbots mix the three pillars of marketing, sales, and service because they are designed to assist the customer in the most beneficial way. During a chatbot conversation, producing a service ticket might be as equally likely as suggesting a product or service to satisfy the customer's need. In the basic working principle of chatbots, the intent of the customer request is determined by a specially designed AI algorithm. Based on the discovered intents, the chatbot dialogue is managed and processed. Usually, chatbots also possess socializing small talk and response delay modules to make the conversation more of a real-world experience.

11.2.2.2 Product Recommendations

Product recommendations are an especially exciting field within robo-advisory because they have a direct impact on the topline of the robo-advisory service. In a provision-based business model such as portfolio management services, regular customer actions are an important revenue driver. At the same time, customers need to be taken care of and be ensured on a regular basis that their investment is in good hands. This provides a good starting position for event-based marketing actions and production recommendations.

Customers who show a high affinity for the happenings in the financial market are interesting from a marketing perspective. Scheduled financial events can be used to promote products, for example, to offer a hedging product in a risk-mitigation scenario.

For example: A customer holds stocks of company XYZ worth USD 10,000 in her portfolio. Quarterly financial results are going to be disclosed for this company in 10 days, and recent quarterly financial reports regularly led to a price jump of 5% the day after the reporting day. Based on these historic findings, USD 500 are potentially at risk. Potential robo-advisory action: A put option is being offered to the customer in order to protect against possible price jumps.

The role of AI in this scenario is manifold. First, customers who show an actual desire to be approached by event-driven marketing actions need to be identified. Approaching all customers would quickly lead to overcommunication, especially when a high number of market events is considered. Consequently, the brute-force way is typically not an option. Second, a product-event match must be found. AI plays a crucial role in this matching process. Typically, the upcoming trigger events' characteristics need to be understood correctly, and well-matching products (e.g. a product for risk mitigation) need to be determined automatically.

11.2.3 Matching Customers to Portfolios

The process of matching customers to portfolios is an important task in robo-advisory, and AI is a supportive factor here. Portfolios of financial assets have varying risk and return levels, and they distinguish themselves in some important characteristics, such as asset class, country/currency exposures, and in the special case of equity investments, industry sectors or other stock attributes on which customers might look for investment opportunities. Income, wealth, and the age of the investor, hence the remaining investment time until retirement, are other critical factors that need to be considered for a proper individual portfolio construction.

Consider two examples:

Example 1: A woman in her mid-20s wants to start putting money in an investment vehicle. She is a supporter of actions against climate change and has an average income but no significant savings. She lives in Europe and does not support Asian environmental politics. For her, a cost-averaging investment approach in a stock portfolio consisting of green US and European companies might be a smart way to (a) be in line with her investment preferences and (b) use the higher long-term return expectations from a stock portfolio. Cost averaging matches her financial situation well.

Example 2: A woman in her late 50s inherits a significant sum of money. She mainly wants to protect this money against market losses and inflation, but she does not want to be exposed to the risk of market fluctuations. To her, the situation looks much different. A multi-asset portfolio consisting of (inflation-protected) high-grade bonds to secure the amount of money mixed with little volatile blue-chip stocks for protecting against inflation might be an answer.

It is a challenging task for human investment advisors to both provide the right ways of investing and accompany investors on a regular basis with their changing investment needs. AI can pre-select a range of well-matching investment opportunities based on the investment needs and preferences of each individual customer. Such an AI system must be skilled in basic investment theory and develop an understanding of which investment opportunities resonate well with certain types of customers. Through this, investment suggestions are generated that are a) economically reasonable and b) appealing to individual customers. One common approach to learn about a customer's investment preferences and economic situation is to guide the person through a set of initial questions. This can help to determine risk appetite, monetary liquidity, and investment goals. Such a service typically falls into the category of investment advice and is a regulated financial service in most jurisdictions. Depending on the jurisdiction, AI might play a role in this matching process or not.

11.3 AI Supporting Portfolio Management

Predicting where financial markets are developing has kept researchers and financial markets participants busy for decades. As it turns out, financial markets tend to show some predictability, especially when today's strong analytics capabilities are applied to financial data from a decade ago (Fischer and Krauss 2018). It is, however, notoriously difficult to tap these sources of predictability, even for experienced researches.

The roots of quantitative finance date back to the beginning of the twentieth century, to Louis Bachelier, a French mathematician, who applied the Brownian motion theory in his PhD thesis to evaluate stock options. Investigating the random-walk behavior of financial products was the starting point to develop a mathematical view of financial processes that led to achievements like the Black–Scholes formula for options pricing, or the Markowitz and later the Capital Asset Pricing Model (CAPM) to manage portfolios in a risk-return adjusted manner. By utilizing stochastic calculus, quantitative finance is used to calculate the fair value of assets and to estimate future risk and return expectations. One famous approach for managing a portfolio of stocks is the three-factor model by Fama and French (1993). In addition to market risk, as discussed in the CAPM, market capitalization and book-to-market ratios are considered to produce buy and sell decisions for each single stock in a portfolio. This approach received a further modification in 2015 by Fama and French, and a five-factor model was proposed. This shows that the quantitative approach to producing profitable financial portfolios is still an actively researched field.

Technical chart analysis is another way how buy and sell signals are generated with quantitative methods. But this approach should not be confused with quantitative finance as described above. In a technical chart analysis, trading rules and trading signals are deducted with the help of technical indicators. An example is the "death cross", denoting a situation when the 50-day moving average price line drops below the 200-day moving average price line. Situations like these only occur quite rarely (e.g. 48 times since 1928 for the S&P 500 Index). From a statistician's viewpoint, quantitative trading rules, such as that mentioned above, typically miss statistical significance. This means that these rules do not possess a sufficiently high degree of reliable forecasting power.

The quantitative finance approach and AI share the idea of predicting future asset prices with mathematical modeling techniques. The application of machine learning is in many ways a natural advancement of the ideas already existing in quantitative finance. Consequently, knowledge about the underpinnings of quantitative finance like the random-walk theory, stationarity requirements of time series, stylized facts of financial time series, to name just a few, are valuable to AI researchers and AI portfolio managers as well. In contrast to classical quantitative approaches, AI and machine learning approaches facilitate a much broader field of explaining variables, such as news comments and sentiments on stocks. Additionally, they rely on more advanced statistical models (such as neural networks in contrast to linear regression) to produce price forecasts.

Today, in the era of big data and advanced analytics, it is even more worthy to investigate this field by applying novel modeling approaches and tapping new data sources. However, new challenges occur and making money by simply throwing powerful algorithms at data has turned out to be not a successful approach in the development of investment strategies. Besides generating alpha (alpha is a common metric for determining the profitability of an investment scheme compared to a benchmark, e.g. a stock market index) with the application of AI, there are other ways in which AI supports the investment process. Especially, when AI is compared to a human investor, some points are worth considering.

- Humans are not free from cognitive biases and are unconscientiously driven by emotions. Behavioral economist and Nobel laureate Daniel Kahneman has discovered and investigated many of these biases, showing how irrationally and illogically people behave in certain situations (Kahneman 2011).
- Decisions driven by "gut feeling" are excluded from the investment process when AI is involved.
- AI typically stimulates process automation and leads to a cleaner and better structured information flow, which in turn increases transparency and traceability.

However, AI is by no means the ideal version of an investor and new issues arise, such as:

- Information may be biased or censored, and an AI algorithm will consequently only learn what it "sees". For example, US stock market prices experienced an extended period of growth (bull market) between 2008 and 2017. An AI that is designed on this period has no understanding of a declining (bear) market.
- An AI algorithm cannot detect correlations in such a way that the induced causational relation really applies. One example is an image-classification AI that is trained to classify images showing wolves and huskies. As it turns out, wolves are often photographed with a snowy background. Consequently, the AI will classify an animal as "wolf" if the background of the image is white. This inference is of course nonsense and in the field of investing money highly dangerous. However, humans are by no means protected from making this mistake but at least have the possibility to question their own decisions on a critical-thinking level.
- Spurious correlations: The strength of an AI algorithm is its capability to digest huge amounts of data, which also becomes its weakness: Correlations may appear at random, and this effect becomes stronger with a growing number of information variables.

A cooperative working model between the human investor and an AI-driven suggestion engine looks very reasonable and combines the strengths of these two approaches. This means, however, that an AI needs to explain itself and pure black-box methodologies do not fall into this category.

There are well-known approaches today that "open" the black boxes of AI algorithms and explain in visual terms the ingredients of the decision-making process (Ribeiro et al. 2016). There is, however, one financial scenario where black-box applications are acceptable: When assets are bought and sold on short time scales (e.g. an average asset-holding period of 5 minutes), a sufficiently high number of profit and loss events are collected quickly. It is then easy to determine with statistical significance if an algorithm works in a profitable manner or not. But this scenario is rather rare for retail-oriented investment vehicles, where rebalancing frequencies are much lower. Even a weekly rebalancing scheme only produces approximately 52 profit and loss events per year, and it is hard to determine the statistically significant correctness of the investment decisions. Hence, pure black-box AI models are not acceptable in this regime.

When developing investment strategies with a multi-parameter AI system, brute-force backtesting (changing parameters until a backtest looks satisfying) is an often applied but certain way to fail. Given the manifold ways for modeling data and additionally considering the many different data sources, some of these combinations will just by coincidence show high profit numbers in a backtest. An ex-post justification of why the best-performing algorithm is also economically reasonable is typically driven by the desire to find supportive explanations and consequently not a reliable way for discovering profitable investment strategies. Hence, some rules of thumb are helpful to prevent this from happening:

1. Start with an investment story that sounds appealing and economically reasonable.
2. Choose zero-parameter systems. These are systems that find their optimal settings by themselves, for example, by parameter optimization. This is helpful because it prevents the strategy developer from playing with the parameters until a good-looking backtest strategy is found.
3. Choose your data wisely. Have an idea why you add a variable to the system and be prepared to explain its behavior from an economic perspective.

4. Always put aside some of the data before the modeling process starts, and these are only unboxed when a satisfying investment model is found.

Apart from finding profitable investment strategies with the help of AI, it might also be used to mitigate risk or reach other valuable goals, such as constructing green portfolios or portfolios that are free from child labor. One key element for achieving these other goals is by tapping alternative data sources, such as text data. Relevant texts can be professional news, as well as information from Twitter or trader forums that have a less curated character.

One stock market example is to blacklist or even short-sell stocks of companies with a bad news sentiment and double down on those with an extraordinary good news sentiment. In a first step, one or more news sources need to be tapped and relevant news articles are extracted. A text sentiment algorithm is used to process these texts, and a sentiment time series is constructed for each stock. Based on the idea that a bad sentiment correlates with a bad company outlook, stocks are either blacklisted or purchased.

Such a sentiment-driven approach can be combined with purchasing decisions from an AI model that generates trading signals based on pure market data. In such a scenario, it would play the role of a risk-mitigating overlay, where stocks with a bad sentiment are removed from the portfolio of stocks. A wide range of data providers promise to deliver such signals in a ready-to-digest format.

Forecasting the stock market is an ongoing field of research for both academia and financial services companies. These research efforts are driven by a constant stream of technological advancements and the desire to find new ways to outperform other market participants.

11.4 The Way Forward

AI systems are developing rapidly with no end in sight. These innovations are strongly pushed by billion-dollar research investments from software companies like Microsoft, Google, SAP, and more. This is especially true for systems that have strong automation and process digitalization aspects, such as chatbots.

Disruptive innovation through AI will be mostly noticeable on the customer end. Here it supports customer interactions, onboarding processes, and service communications to run smoothly, quickly, and with a clear customer focus. AI will be a key driver in scaling customer processes and services and will have a dominating impact on the top and bottom line of a robo-advisory service.

Even though highly interesting from an intellectual perspective, investment portfolios run by AI will be more of a cost versus benefit discussion and less of an alpha-generation question. Hence, some investment decisions and processes may well be entirely replaced by AI systems, given that the benefit of introducing these systems outweigh the significant cost of installing them. However, one question that has been around for decades is how safe consumers feel when their money is managed by algorithms instead of humans. This is certainly a question of time, as people feel safe in aircrafts without questioning the auto pilot. AI systems will most certainly not replace skilled professionals but rather aid in their work.

For the highly digital business model of robo-advisory, having AI at the core of the strategic development is a key success ingredient. This is true not only from a customer satisfaction perspective but also from a revenue and operational cost perspective. Especially for early-stage robo-advisory companies, a structured and scalable data strategy is a key component for a later profitable utilization of AI.

The bottom line is that robo-advisory is a highly attractive playing field for AI systems. The strongest impact of AI systems will be felt on the customer-management side and much less on the asset-management side.

References

Association for Computing Machinery. 2018. *Fathers of the Deep Learning Revolution Receive ACM A.M. Turing Award*. Last Accessed 20 March 2020. https://awards.acm.org/about/2018-turing.

Breiman, Leo. 2001. "Random Forests". *Machine Learning* 45(1): 5–32. https://doi.org/10.1023/A:1010933404324.

della Cava, Marco. 2016. *Microsoft CEO Nadella: 'Bots are the new apps'*. Last Accessed 13 September 2019. https://eu.usatoday.com/story/tech/news/2016/3/3/microsof-ceo-nadella-bots-new-apps/82431672/.

Domingos, Pedro. 2012. "A Few Useful Things to Know About Machine Learning". *Communication of the ACM* 55(10): 78–87. https://doi.org/10.1145/2347736.2347755.

Durkin, John. 2002. "History and Applications". In *Expert Systems*, ed. Cornelius T. Leondes, 1–22. New York: Academic Press. https://doi.org/10.1016/B978-012443880-4/50045-4.

Fama, Eugene F., and Kenneth R. French. 1993. "Common Risk Factors in the Returns on Stocks and Bonds". In *Journal of Financial Economics* 33(1): 3–56. https://doi.org/10.1016/0304-405X(93)90023-5.

Fischer, Thomas, and Christopher Krauss. "Deep Learning with Long Short-Term Memory Networks for Financial Market Predictions". *European Journal of Operational Research* 270(2): 654–669. https://doi.org/10.1016/j.ejor.2017.11.054.

Giles, Timothy D. 2016. "Aristotle Writing Science: An Application of His Theory". *Journal of Technical Writing and Communication* 46(1): 83– 104. https://doi.org/10.1177/0047281615600633

Hastie, Trevor, Robert Tibshirani, and Jerome Friedman. *The Elements of Statistical Learning: Data Mining, Inference and Prediction*. Springer Series in Statistics. New York: Springer. https://doi.org/10.1007/978-0-387-84858-7.

Hernandez, Pedro. 2017. *Microsoft Officially Releases Azure Chatbot Service*. Last Accessed 13 September 2019. https://www.eweek.com/enterprise-apps/microsoft-officially-launches-ai-bot-creation-services-for-developers.

Kahneman, Daniel. 2011. *Thinking, Fast and Slow*. New York: Farrar Straus and Giroux.

Ribeiro, Marco T., Sameer Singh, and Carlos Guestrin. 2016. " "Why Should I Trust You?": Explaining the Predictions of Any Classifier". *Machine Learning* 97–101. https://doi.org/10.18653/v1/N16-3020.

Seaborn, Katie, and Deborah I. Fels. 2015. "Gamification in Theory and Action: A Survey". *International Journal of Human-Computer Studies* 74: 14–31. https://doi.org/10.1016/j.ijhcs.2014.09.006.

CHAPTER 12

What Role Do Social Media Play for Robo-Advisors?

Ana-Maria Climescu

Contents

A.-M. Climescu (✉)
Hauck & Aufhaeuser, Frankfurt am Main, Germany
e-mail: ana-maria.climescu@hauck-aufhaeuser.com

© The Author(s) 2021

P. Scholz (ed.), *Robo-Advisory*, Palgrave Studies in Financial Services Technology, https://doi.org/10.1007/978-3-030-40818-3_12

12.1 Introduction

Looking at the current number of users, it becomes clear that social media[1] today play an important role both privately and professionally. Their rise is mainly attributable to the increasing use of the internet in general (Statista 2019a) and smartphones and other mobile devices in particular (Statista 2019b). In January 2019 Facebook (FB) had approximately 2.27 billion active users per month worldwide, making it the largest social network in the world (Statista 2019c). Second place goes to YouTube with 1.9 billion users, which is owned by Google (Statista 2019c). WhatsApp follows in third place with 1.5 billion active users per month (Statista 2019c). The overall distribution of worldwide use of social media channels is summarized in Fig. 12.1: This presentation of worldwide user figures shows the relevance of social media in people's everyday lives. Nevertheless, there are of course large differences between individual countries. In the USA, for example, Facebook (68% usage) and YouTube (73%) are the leading social media in 2018 (Sandmeier 2018). In addition, Instagram is growing rapidly in the USA, while Facebook user numbers are stagnating (Sandmeier 2018). In Asia, on the other hand, WeChat and QQ dominate the market (Mao 2018). In Germany, Facebook (31%) Instagram (15%), and Snapchat (9%) are again in the lead in 2018 (ARD & ZDF 2018). The age structure is also interesting here. For example, Instagram and Snapchat are preferred in the 14–19 age group, while Facebook is increasingly used in the 20–29 age group (ARD & ZDF 2018).

This high number of users and frequency has led to social media already being seen as an established marketing instrument—not least because the business models are largely dependent on advertising revenues. And so almost every company today is expected to have social media channels at its disposal in order to win new customers and retain those already acquired (Kontor4 2019). In the context of robo-advisory, three relevant questions can be derived for this section:

1. Have social media favored the emergence of robo-advisors?
2. What role do social media in general play for robo-advisors?
3. How can social media support robo-advisors?

[1] Social media in this context are understood as communication platforms such as Facebook, Instagram, Twitter, YouTube, LinkedIn, Xing, and Pinterest, which enable users to share content digitally and interact with other users via an online application/app. In addition there are messenger services like WhatsApp, Snapchat, FB Messenger, Skype, and iMessage. This definition is based on the lexicon of Gruenderszene, but has been extended to include those channels that are relevant to this section (Gründerszene 2019).

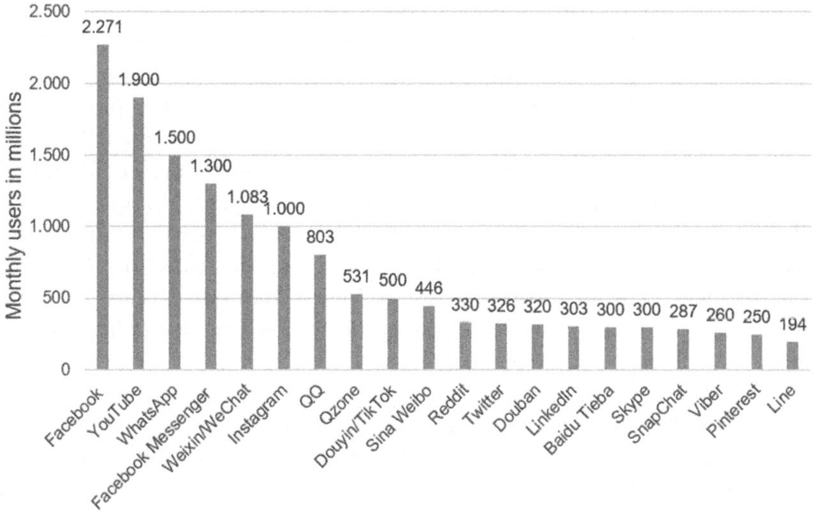

Fig. 12.1 Active users of social media worldwide (Own illustration based on Kontor4 2019)

12.2 Social Media vs. Robo-Advisor

To answer the first question, we will first take a look at the evolution of social media in comparison to robo-advisors to find out whether there has been temporal parallelism or interdependence. Although there were already various platforms in the 1990s that sought to connect people and enable the direct exchange of messages, it was LinkedIn that laid the foundation for today's social and digital media landscape in 2003 (Kroker 2018). One year later, in 2004, Facebook came onto the market (Kroker 2018). This was followed by YouTube in 2005 and MySpace and Twitter in 2006 (Kroker 2018). In 2008 FB launched its Messenger. Instagram did not follow until 2010 (Kroker 2018). Since then, the number of users has been rising constantly, as already mentioned (Kontor4 2019). The comparison with the evolution of robo-advisory is intriguing.

The chronological development of robo-advisors has been traced in a study by Becchi et al. (2018, p. 3). Accordingly, their emergence can be traced back to 2008, with the first services from Wealthfront and Betterment, among others. In 2010, Bank of America and Merrill Lynch Merrill

Edge introduced a robo-advisor to optimize investments for clients. Two years later, Australia and New Zealand Banking Group Limited (ANZ) began using artificial intelligence (AI) technology to better understand the customer. In 2015, Blackrock acquired a digital advisory platform to improve investment decisions. Meanwhile, in 2017, Betterment was one of the first robo-advisors to announce that it would develop its digital business through client advisors in the direction of a hybrid business model. Robo-advisors have therefore experienced a massive growth in the period from 2008 to today, which is also reflected in searches for this term on the net.[2] The first robo-advisor followed about five years after the first social medium. There is therefore no immediate chronological parallelism. But the close chronological succession in which the emergence of robo-advisory follows social media suggests an interdependence. Indeed, between 2005 and 2010 in general, there was a major technological advance that affected many industries. The first iPhone (2007) is one of many technological developments that originated in the same time span and have digitized the lives of a large number of people. With smartphones and social media, people's user behavior began to change step by step. This was accompanied by an increased desire for digital, location-independent services and access to products. The new technological possibilities have had an extreme impact on user behavior and thus on customer wishes; and have opened the door to new digital business models (Ernst & Young 2018, p. 42).

In the same period (2008), however, there was another significant change for banking services that has also shaped the behavior of bank customers worldwide: the financial crisis. In addition to heavy economic losses, it was above all the loss of confidence in existing institutions, accompanied by increasing cost pressure on financial service providers, that paved the way for the emergence of new alternatives (Ernst & Young 2018, p. 42). Against this background, banks were forced to address on the one hand the options for automating their own processes and on the other hand the greater skepticism on the part of customers, coupled with new customer requirements. This put existing business models to the test, while some existing financial service providers, together with some new ones, seized the opportunity offered by the latest technologies (Vogel

[2] An analysis of the terms *fintech* (as a general umbrella term) and *robo-advisor* using Google Trends shows an increase in public interest from 2014 (in the US and Germany).

2016). The emergence of robo-advisors and the digitalization of existing banking processes, such as creating an investment proposal, is therefore due to three factors: the financial crisis, technological opportunities and the digital affinity of customers. The last of these factors was promoted by social media, which, along with e-commerce, have the greatest direct influence on day-to-day user behavior. In addition, they were used by the robo-advisory market as a new digital marketing platform that can accompany users along their entire customer journey—from acquisition to after-sales service.

12.3 Robo-Advisor Feat. Social Media

Social media is much more than a simple extension of marketing funnels and has a significant influence on the business models of many companies. Particularly after the financial crisis, social media offered a place for those affected and interested users to interact, which created a whole new level of transparency (Planung & Analyse 2011). This has opened up new opportunities for digital business models such as robo-advisors. In concrete terms, four major areas of influence of social media can be identified for robo-advisors: Marketer/enabler (Becchi et al. 2018, p. 4), communication channel, data source and (on a broader level) even strategy. According to a study by EY, social media platforms are considered "enablers", that is, pioneers, of robo-advisors which firstly serve to cultivate customer relationships and secondly increase the reach of the brand (Becchi et al. 2018, p. 4). They provide a space in which corporate content can be made publicly available. In addition, they function as a platform on which an open exchange of experience can take place and a relevant community can be established.[3] Industry-relevant influencers can help strengthen brand trust and increase reach within the community (Mindruta 2016). Furthermore, providers can communicate directly with users, for example, via chat, which promotes customer acquisition and retention (Becchi et al. 2018, p. 5). Social media thus also function as a communication channel and offer direct access to the people behind the digital service, which is particularly valuable for digital investment models without constant advice. Direct contact with users and customers is also elementary from the company's

[3] The term "fintech marketing" was coined in this context, which describes the marketing of digital financial service providers and banks with the help of social media. The goal: mutual interaction through contributions and thus support in building a community (Finleap 2017).

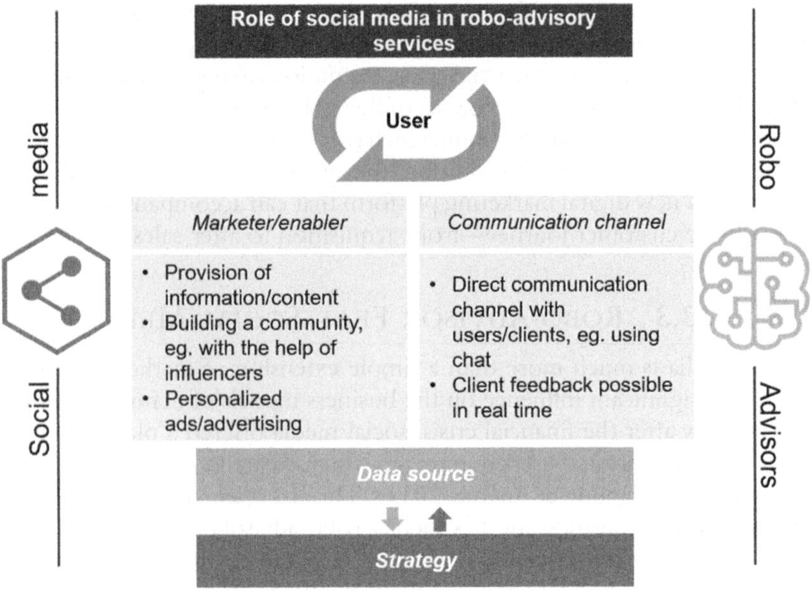

Fig. 12.2 Areas of influence of social media on robo-advisors (authors' own image)

point of view: feedback and evaluations from users can reveal valuable improvement potential in their own products (Fig. 12.2).

Finally, social media offers opportunities for effective (social media enables accurate tracking) and personalized marketing (user behavior on social media allows conclusions to be drawn about other needs of a (potential) customer) and thus to attract customers and increase revenue (Becchi et al. 2018, p. 4). Advertising created on the basis of personal characteristics increases the relevance of the offer for the customer and reduces wastage on the part of the provider.[4] If you look beyond robo-

[4] Thus, FB, Instagram, YouTube and LinkedIn, for example, enable personalized advertising to be displayed that takes factors such as demographics, purchasing preferences and friends into account, so that advertising banners are shown only to those users who might be interested in the respective company or product/service according to their personal characteristics.

advisors, you can already find impressive examples of this. For instance, American Express has given its customers the opportunity to link their Facebook or Foursquare accounts to their Amex cards so that they can be provided with personal offers based on data from these accounts (Hamilton 2016). As a result, social media no longer simply offer the possibility of distributing content, but can also act as an interface with the producer through the use of AI in order to influence the buying behavior of customers (Finleap 2017). The primary focus is on big data evaluations: social media are a rich source for this, as FB (in possession of a European banking license) has already proven impressively. Today users voluntarily provide companies like FB with a great deal of personal information that can be used for targeted marketing/personalized ads if analyzed correctly.

However, the use of AI is particularly promising in this context, as it could help robo-advisors, among others, to exploit the full potential of social media beyond their traditional marketing function. What this means in concrete terms will be explained briefly in the following, using robo-advisors as an example. Until now, robo-advisors have exclusively used the existing technology and functions that social media offer them, for example, as a publication platform, an appointment agreement tool or a chatbot. The turning point in terms of efficiency and customer satisfaction would now come if a provider were able to use AI to collect and process the personal data of users of its own social media channels and forward it in real time to the product platform, which then adapts itself automatically to customer needs. For example, market trends for certain investment products would lead to an immediate adjustment of the product portfolio. Decisions about costs or marketing presence would then no longer be made by the company itself during ongoing operations, but would be handled by an integrated AI platform. Through the use of AI, social media can thus point the way for the entire strategy of robo-advisors in future (FinTech Futures 2018).

12.4 Social Advisory

In general, some interesting approaches for the current, but above all for the future significance of social media for robo-advisory can now be derived from the previous discussion. In relation to the question as to whether social media have favored the emergence of robo-advisors and what their overall role is for the robo-advisory market, the following conclusions can be drawn:

1. The use of social media continues to grow worldwide, with regional and continental differences in relation to preferred media. While Facebook, YouTube and Instagram dominate the market in the USA, Facebook, Instagram and Snapchat lead the market in terms of user numbers in Germany, for example. For the robo-advisor market, this means that media must be evaluated on a regional basis with regard to their use. The age of the target group for the respective product also plays a crucial role.
2. Looking at the historical development of social media in comparison to that of robo-advisors, it is also evident that social media are about five years ahead of the first robo-advisor. This goes hand in hand with increasing digital affinity, with digital consumption and overall attraction to digital technology now increasingly taking hold in people's lives. This different world is increasingly reflected in related customer demand and digital expectations of companies in general and financial services in particular.
3. In relation to the role of social media for robo-advisors, this results in a total of four central areas of application: marketer/enabler, communication channel, data source and strategy. Social media enable robo-advisor providers to maintain their customer relationships through direct contact and increase the reach of their brand. As a communication channel, social media can also enable personalized offers and targeted advertising to be provided on the basis of the data collected. This not only provides valuable insights for the strategy (products in demand and pricing), but also potentially allows automated adjustments to the product platform via AI.

Against this background, the following points can be considered going forward with regard to the scope for the use of social media to support robo-advisory models:

12.4.1 Social Media as Marketing and Sales Support

Social media can serve as an interface between the end consumer and the digital financial services provider. In concrete terms, this means that they are used as a platform to communicate (one-to-one-to-many)—that is, via chat, telephone, in public or in private. For example, FB already offers a scheduling function that companies can use to arrange meetings with potential customers. For many e-commerce providers, initiating business via social media has already become part of everyday life.

12.4.2 Social Media as an Alternative to Robo-Advisory

And this potential is now leading to a consideration of why social media should not offer more financial services. Not only do they have the necessary data, such as place of residence, education or profession, to present tailor-made offers to customers, but they also have the necessary resources and know-how to evaluate and use these data volumes. Since 2016 Facebook has had a license for money services with the Irish Central Bank (Degel 2016). It was not until mid-2019 that the group announced its plans for its new payment service libra. The new platform, in which PayPal and Visa, among others, would like to participate, will be based on blockchain technology (Schuler 2019). It has also been reported on occasion that Snapchat is working on a robo-advisor (Wolff-Mann 2016). The reasons for the failure to exploit this potential to date probably lie in the regulations for financial service providers and in the complexity of banking products. WeChat, on the other hand, has already implemented a payment function (Taylor 2019). However, more complex bank transactions such as financial investments are not possible in China either.

12.4.3 Social Media as Potential Cooperation Partner for Robo-Advisors

Nevertheless, these tendencies and attempts on the part of social media to enter the banking business lead to the conclusion that a cooperation between social media and robo-advisors, that is, banking services in the field of investment, could certainly be a strategic consideration. The technical course for this has already largely been set by the social media and modern robo-advisors, in the form of social-advisors. For both business models, the next step is to evaluate the potential of such a cooperation from a business and strategic point of view using concrete case studies.

References

ARD & ZDF. 2018. *Onlinecommunities – Nutzung von Onlinecommunitys 2018*. Last Accessed: 7 September 2019. http://www.ard-zdf-onlinestudie.de/whatsapponlinecommunities/.

Becchi, Scott M., Ugur Hamaloglu, Taroon Aggarwal, and Samit Panchal. 2018. *The Evolution of Robo-advisors and Advisor 2.0 Model: The Future of Investment Management and Financial Advisory*. Ernst & Young Global Limited.

Degel, Joachim. 2016. *Facebook Hat Jetzt eine E-Bank-Lizenz für die EU*. Last Accessed 17 March 2019. http://www.bgk-p.de/de/blog/posts/Facebook-hat-jetzt-eine-E-Bank-Lizenz-fuer-die-EU.php.

Ernst & Young. 2018. *EY Banken-barometer 2018 – 10 Jahre nach der Finanzkrise – Aufbruchstimmung?* Tech. rep. https://www.de.ey.com/Publication/vwLUAssets/ey-bankenbarometer-mai-2018/\protect\LY1\textdollarFILE/ey-bankenbarometer-2018.pdf.

Finleap. 2017. *Fintech Trends: Social Media*. Last Accessed 17 March 2019. https://www.finleap.com/insights/fintech-trends-social-media/.

FinTech Futures. 2018. *Social Media and Fintech – What Kind of Relationship?* Last Accessed 17 March 2019. https://www.bankingtech.com/2018/05/social-media-and-fintech-what-kind-of-relationship/.

Gründerszene. 2019. *Social Network*. Last Accessed 17 March 2019. https://www.gruenderszene.de/lexikon/begriffe/social-network?interstitial.

Hamilton, Josch. 2016. *How Social Media is Changing Fintech and Other Startups*. Last Accessed 17 March 2019. https://cloudtweaks.com/2016/12/social-media-changing-fintech-startups/.

Kontor4. 2019. *Social Media 2019: Aktuelle Nutzerzahlen*. Last Accessed 17 March 2019. https://www.kontor4.de/beitrag/aktuelle-social-media-nutzerzahlen.html.

Kroker, Michael. 2018. *Die Social-Media-Zeitleiste: Von der LinkedIn-Gründung 2002 bis zum IGTVStart 2018*. Last Accessed 17 March 2019. https://blog.wiwo.de/look-at-it/2018/07/16/die-social-media-zeitleiste-von-der-linkedin-gruendung-2002-bis-zum-igtv-start-2018/.

Mao, Bianca R. 2018. *Asian Social Media: 5 Networks You Should Know About*. Last Accessed 7 September 2019. https://www.synthesio.com/blog/asian-social-media/.

Mindruta, Ruxandra. 2016. *Was ist das Geheimnis? Eine Einführung in das Influencer Marketing – Markenreichweite erhöhen. Glaubwürdigkeit verstärken. Viele Euros einsparen. Dies sind nur einige Vorteile von effizienten Influencer Marketing Strategien*. Last Accessed 17 March 2019. https://www.brandwatch.com/de/blog/was-ist-das-geheimnis-eine-einfuehrung-in-das-influencer-marketing/.

Planung & Analyse. 2011. *Social Media thematisiert die Finanzkrise.* https://www.horizont.net/planung-analyse/nachrichten/Social-Media-thematisiert-die-Finanzkrise-148096.

Sandmeier, Michelle. 2018. *Pew Research Center: Social Media in den USA 2018.* Last Accessed 7 September 2019. https://bernet.ch/blog/2018/03/08/pew-research-center-social-media-in-den-usa-2018/.

Schuler, Marcus. 2019. *Holt Facebook die Kryptowährungen aus der Nische?* Last Accessed 17 March 2019. https://www.deutschlandfunk.de/neues-digitalgeld-libra-holt-facebook-die-kryptowaehrungen.684.de.html?dram:article_id=452038.

Statista. 2019. *Anzahl der Internetnutzer weltweit in den Jahren 2005 bis 2017 sowie eine Schätzung für das Jahr 2018.* Last Accessed 17 March 2019. https://de.statista.com/statistik/daten/studie/805920/umfrage/anzahl-der-internetnutzer-weltweit/.

Statista. 2019. *Prognose zum Anteil der Smartphone-Nutzer an den Mobiltelefonnutzern weltweit von 2014 bis 2021.* Last Accessed 17 March 2019. https://de.statista.com/statistik/daten/studie/556616/umfrage/prognose-zum-anteil-der-smartphone-nutzer-an-den-mobiltelefonnutzern-weltweit/.

Statista. 2019. *Ranking der größten sozialen Netzwerke und Messenger nach der Anzahl der monatlich aktiven Nutzer (MAU) im Januar 2019.* Last Accessed 17 March 2019. https://de.statista.com/statistik/daten/studie/181086/umfrage/die-weltweit-groessten-social-networks-nach-anzahl-der-user/.

Taylor, Twain. 2019. *Alipay and Wechat Pay: The Quest for a Cashless China.* Last Accessed 17 March 2019. http://techgenix.com/alipay-and-wechat-pay/.

Vogel, Michaela. 2016. *Social Media-Analyse zur FinTech-Branche - In unserer Social-Media-Analyse warfen wir einen Blick auf die Online-Gespräche rund um FinTechThemen und Unternehmen.* Last Accessed 17 March 2019. https://www.brandwatch.com/de/blog/social-media-analyse-zur-fintech-branche/.

Wolff-Mann, Etan. 2016. *Snapchat Wants to Be Your Next Financial Adviser.* Last Accessed 7 September 2019. https://money.com/snapchat-robo-financial-adviser-investing/.

CHAPTER 13

Success Factors for Robo-Advisory: Now and Then

Madeleine Sander

Contents

M. Sander (✉)
Hauck & Aufhaeuser, Frankfurt am Main, Germany
e-mail: madeleine.sander@hauck-aufhaeuser.com

© The Author(s) 2021

P. Scholz (ed.), *Robo-Advisory*, Palgrave Studies in Financial Services Technology, https://doi.org/10.1007/978-3-030-40818-3_13

13.1 Introduction

After reading all the implementation details, facts, and best practice in the field of robo-advisory—what are the key takeaways to make robos a success story? What are the hurdles that robo-advisory will need to overcome in order to succeed? And what are the crucial benchmarks? To answer these questions, it is worth looking first at the US and thus at the roots of the robo-advisory market. The first robo-advisor, the fintech Betterment, was launched in the US in 2008 at the climax of the financial crisis. The robo-advisory market in the US not only has the longest history but is also currently the global leader in terms of assets under management (AuM); for details on market data see Chap. 1.4. Although the robo-advisory market is still largely in its infancy, the amount of empirical data it has at its disposal is unique on a global scale.

In 2018, the top five robo-advisors accounted for an estimated 70% of assets under management in the US (Friedberg 2018). While Betterment ranked third in terms of assets under management (AUM) in 2018, asset managers and online brokers such as Vanguard and Charles Schwab topped the list with "Vanguard Personal Advisor Services" on first place and "Schwab Intelligent Portfolios" on second place (Friedberg 2018). The involvement of these established financial institutions has made a significant contribution to growth. This is mainly due to the high brand recognition, the trust placed in the brand, and leveraging the cross-selling potential of existing clients. These are the two key factors that give traditional asset managers a competitive edge over fintechs, which are the actual inventors of robo-advisory.

Similar patterns are emerging on the German market. In 2018, the fintech Scalable, in partnership with the direct bank ING Diba, was a clear market leader with a market share of approx. 25% (for more details on market data please refer to Chap. 1.4). Furthermore, the current trend of mergers between established providers and the new robo-advisors will continue; the latest examples include Deutsche Bank with Robin, as well as Hauck & Aufhaeuser Privatbankiers with Zeedin. At the same time, the first signs of market consolidation started to emerge in 2018 with Werthstein discontinuing business operations (Robo-Advisor Portal 2018) and the end of cooperation between Haspa and the fintech Investify (Brummer 2018), as well as the takeover of Vaamo by Moneyfarm (Robo-Advisor Portal 2018). On first glance, the underlying reasons seemed to be quite diverse. In the case of Werthstein, sales fell short of expectations (l.c.), while Moneyfarm and Vaamo intend to use the merger to set up an online asset manager operating throughout Europe (IT Finanzmagazin 2018). In contrast, the Finanz-Szene portal reported that "Haspa's robo plans failed, not least because of internal resistance" (Dohms 2018). Nevertheless, there are two distinct patterns emerging from the cases described. Given the considerable marketing costs for new customer acquisition and the low cross-selling potential of robo-advisory, the robo-advisory market is a scale market. Market participants who fail to generate reach within the first two to three years (e.g. by moving up to the top 10 in their home market), or to leverage synergies with existing business areas and to optimize the growth-related KPIs, will sooner or later be swept up by the consolidation wave. On the other hand, parallel interests may play a central role in the successful implementation of cooperative ventures, as well as the introduction of new products at established institutions, supported by appropriate internal incentives. This can also mean that higher-margin products are sacrificed in favor of lower-margin, progressive products in the interest of sustainable success. This is because the entire global robo-advisory market is still expected to experience immense growth. The consulting firm Roland Berger expects the CAGR of robo-advisors' assets under management to be between 20 and 40% by 2022 (Buess et al. 2018, p. 8).

When considering these growth forecasts, however, the question arises—as with any product—where will this growth come from. Essentially, there are two options: either it will be driven by general growth in the respective market (i.e. the investment volume, for example, through an average increase in private assets) or by pushing out other products (i.e. investment opportunities, such as a self-directed securities account or a direct investment in ETFs or actively managed funds). Another factor to keep in mind when it comes to investing is the fact that people still have large amounts of money and thus potential investment volume sitting in their checking accounts—especially in Germany. Here, assets parked in demand deposits, and cash amounted to EUR 1.6 trillion in the last quarter of 2018, representing an increase of 188% since 1999.[1] Releasing this potential could be the third factor, although experience has shown that it is not that easy. According to a study by JP Morgan Asset Management in cooperation with GfK, only 5% of Germans surveyed switched to more profitable investment products after a decade in a low-interest environment

[1] See TagesgeldVergleich (2019).

(Bradtmöller and Düchting 2018). At the same time, 81% of the women and men surveyed are dissatisfied with the performance of their savings products (Bradtmöller and Düchting 2018).

13.2 Growth Factors Identified

13.2.1 No. 1: Overall Market Growth

While the global markets—measured in AuM or as assets per capita—are expected to grow by 6 to 7% p.a. up to 2022, the fastest growing markets in terms of assets per capita are expected to be in Eastern Europe with 11% p.a. and Central Asia with 10% p.a. until 2022 (Beardsley et al. 2018, p. 11). This has the potential to be an important piece of the puzzle from which all investment products can benefit. The simultaneously expected increase in e-commerce—approx. 12% p.a. up to 2022 in retail online sales in the US and Western Europe versus a growth rate of 3% p.a. in the retail sales as a whole (Bhave et al. 2018, p. 7)—should also favor digital financial solutions such as robo-advisors.

13.2.2 No. 2: Replacing Existing Products

With respect to the second factor, the relative customer benefits of robo-advisors compared to conventional investment opportunities for private clients, whether in the retail or affluent segment—such as buying an actively managed fund or investing in ETFs—point toward a significant potential. At around 0.3% p.a., for example, the ETF fees are relatively attractive and about 1.3% p.a. lower than the costs of active funds.[2] On the other hand, with the conventional ETF, the client participates one-to-one, both in the value gains and in the value declines of the replicated index. ETFs do not offer services such as risk management in the form of ongoing rebalancing in line with the previously determined investor's investment and risk profile or other methods typically used by robo-advisors. Furthermore, it is also debatable how many and which ETFs to pick to achieve a differentiated portfolio. Private clients attempting to invest into ETFs beyond the equities asset class will need specialist knowledge—for example, on the functioning of bonds and correlations

[2] Fonds und ETFs: Aktiv versus passiv (Focus Money No. 28/2017, p. 53).

between asset classes. However, the growth in popularity of ETFs over the last few years has highlighted the importance of transparency and low costs. With a conventional robo-advisor, the prospective client receives an investment proposal for a diversified portfolio with several financial securities such as ETFs, actively managed funds through to individual stocks or certificates in line with the previously determined investment profile. The cheapest robo-advisors in Germany cost 0.6% of the invested amount per year (Test 2018). The growth potential of robo-advisors by replacing existing investment opportunities or in line with existing investment products—ETFs are currently robo-advisors' main investment product—is, therefore, quite substantial.

Nevertheless, some transparency and education is required to come close to the reach of established investment opportunities. For example, the global Google search requests for robo-advisors in 2018 were just under 400,000 vs. 4.4 million for ETFs, at the same time, according to an IPSOS 2018 (Antoniotti et al. 2018, p. 5) study in the US, only 23% of respondents said they were very familiar with robo-advisors, and 5.3% said they were already using robo-advisors (Fig. 13.1). However, compared to 2015, the figures of 11 and 3.8% have increased significantly. A similar survey was conducted by Ebase in early 2019 for the German market. Roughly, only every fifth person knew what a "robo-advisor" was (Nicolaisen 2019). After all, the figure was just over 28% for men and 46% for individuals with a net income of at least EUR 4000 (Fig. 13.2). At the same time, the projections for the future are quite positive. For example, across all respondent groups, 42% rate the likelihood of using a robo-advisor to invest over the next 12 months as medium to high. At 46%, women are still ahead of men with 40%. The respondents considered the factors of transparency, low fees, and traceability of the investment strategy to be most important (Fig. 13.3).

13.2.3 No. 3: Attracting Previously "Uninvested" People to the Financial Markets

The third lever is certainly the most challenging, although there is plenty of potential behind it, and rising inflation coupled with continued low interest rates in Germany in 2018, for example, has amplified the pain point for private customers who park their cash in a checking account. The question, therefore, is what would it take for these largely inexperienced investors to choose a robo-advisor?

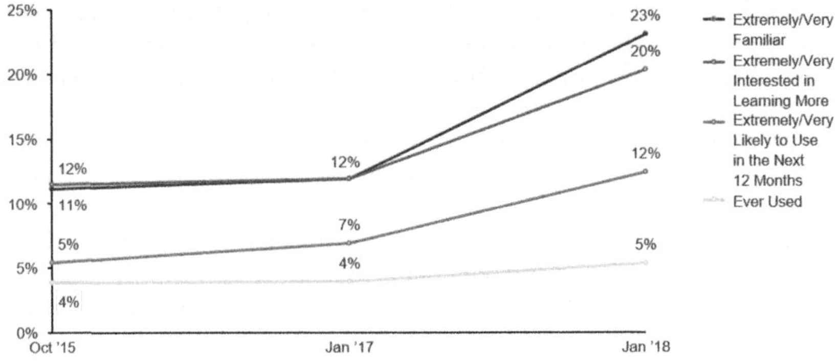

Fig. 13.1 IAI Q1 2018 USA Barometer on awareness of robo-advisor based on Antoniotti et al. (2018)

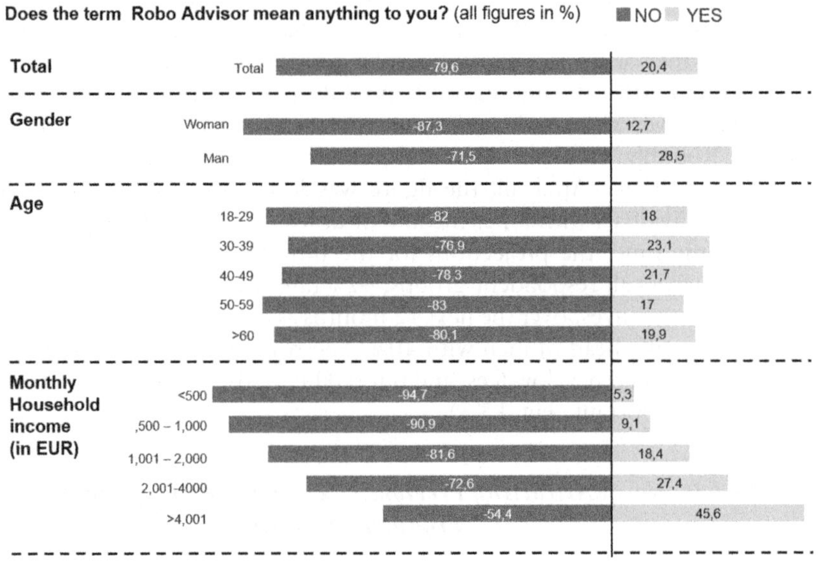

Fig. 13.2 Ebase survey Q1 2019 on the awareness of the term robo-advisor based on Nicolaisen (2019)

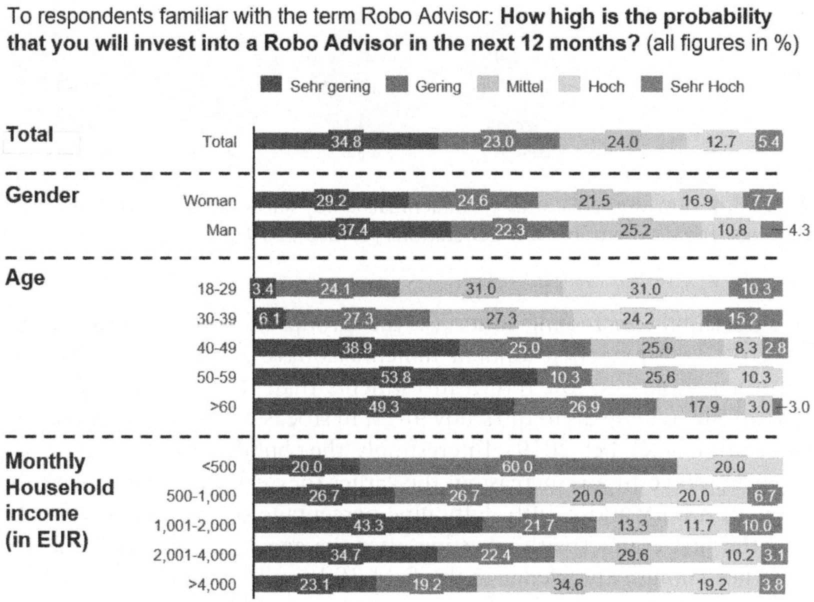

Fig. 13.3 Ebase survey Q1 2019 on the likelihood of using a robo-advisor to invest in the next 12 months Nicolaisen (2019)

According to a representative Forsa survey in 2018, 43% of respondents in Germany stated that they parked their money in a savings account. On the other hand, according to the Forsa survey, 28% of those surveyed indicated that they invest in investment funds and 24% in equities. The results are even lower for women: 60% even rule out ever investing their money in stocks or funds (Wenig 2018). One of the main reasons appears to be their lack of knowledge about investing (l.c.). For example, 48% admitted that they were not well informed about the possibilities of investing money, and "(...) one in four respondents (24%) admitted[,] they had little or no knowledge about investment opportunities," (l.c.) while in a survey conducted by the Association of German Banks (Bundesverband deutscher Banken), only 44% stated that they knew what an investment fund was (Verbrauchermagazin-Redaktion 2017).

Furthermore, a study from Investopedia in cooperation with Chirp Research shows that the reservation against stock market investments

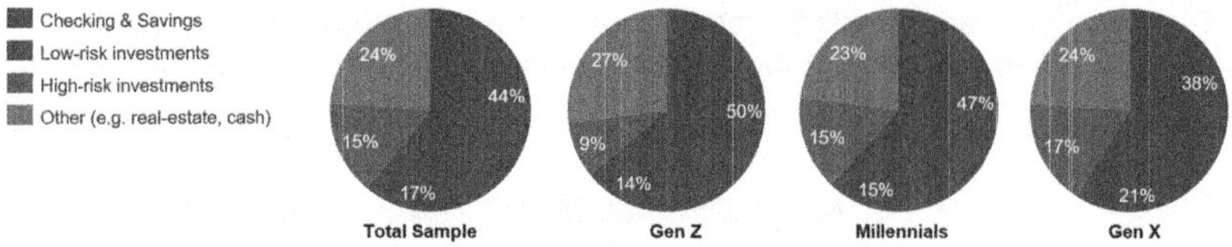

Fig. 13.4 Investopedia affluent millennials study shows the wealth distribution amongst the different investor generations (Gobell 2019)

is surprisingly increasing amongst the group of younger generations. Figure 13.4 shows that affluent millennials have a higher probability to put their money in low return investments than the generation X. Only 34% of the millennials in the study invest in stocks compared to 47% of the generation X (Gobell 2019). Interestingly, the confidence of the millennials regarding investing is increasing, the earlier they came in contact with the topic of money and wealth. Fifty nine percentage of the millennials, who had the first contact with investing after the age of 21, feel that they do have not enough knowledge in this field. If the millennials in the study had contact with investing before the age of 15, only 12% have the impression of being under-educated in the subject (Gobell 2019). A targeted approach that addresses the individual needs of this target group can, in principle, have a positive effect on the growth of all existing financial market products. At the same time, however, the answers cited above suggest that investment opportunities with an edge are those which are easy to understand and/or make the process as straightforward as possible for the investor. The first step when investing with the help of a robo-advisor is to produce a personal risk profile for the investor based on their individual goals, risk appetite, and experience. With robo-advisors, investors do not have to manage their portfolio themselves as robo-advisors pick the stocks for them and do the ongoing rebalancing as is the case with conventional, analog discretionary portfolio management, which has so far only been typically available to wealthy clients with at least EUR 0.5 million to invest. These factors, as well as the points mentioned under growth factor 2, point to a significant potential for robo-advisors. To unlock this potential, a targeted expansion of the existing offering is required.

13.3 Success Factors for Achieving the Identified Potentials and Increasing the Market Penetration of Robo-Advisors

The potential behind the highlighted factors is quite immense and supports the growth forecasts of some well-known management consultants and statistics portals. What accompanying activities could help to unlock this potential? What can be done to address the cited arguments such as the lack of reach, transparency, financial market knowledge, and the perceived complexity? Six possible approaches are outlined below:

13.3.1 Extending the Reach

Cooperation with established partners and brands can make a significant contribution here. These can involve cooperation within the financial sector (such as the cooperation between Scalable and ING Diba) as well as innovative, cross-sector ventures (for example, Ikea partnering with Sonos,[3] the men's fashion designer Virgil Abhol of Louis Vuitton (Samann 2019) or Adidas (Baur 2018)). In this respect, it is key to seek win-win partnerships offering broadly equal benefits to both parties. This is especially the case if both partners offer something that the other party lacks. In addition to the much-discussed financial industry collaboration between fintechs (a robo-advisor fintech, in this instance) and established financial market participants with existing reach in the relevant target customer groups, robo-advisor providers should also consider more creative, cross-sector approaches (think Ikea), especially in the era of open banking. To realize the identified cooperation potentials, it is also very helpful if the cooperation partners have compatible corporate cultures and the key individuals from both sides involved in the project get on well, even on a personal level. Buzz campaigns are another important element, illustrated by Red Bull with its event marketing campaigns featuring extreme sports (e.g. "Red Bull Stratos—Felix Baumgartner's Space Jump" Sampiero 2013 or "Red Bull Crashed Ice" Red Bull 2019). In addition, even in the age of digitalization, and maybe because of it, personal networks can play a pivotal role in the promotion of a product (especially in the form of word-of-mouth advertising).

[3] See Computer Bild (2019).

13.3.2 *Improving Financial Literacy*

This is an extremely important and at the same time fundamental issue for all investment products. Today, it is still not uncommon to go through school and even further education, including universities, without really coming across the topic of financial education. Where is the financial knowledge supposed to come from if it is not acquired privately? The same applies to industries such as law or medicine. While experts are usually trusted here, trust in the financial industry has suffered greatly in recent years. One effective remedy here is to offer help with self-learning to build up some basic knowledge. This calls for the involvement of various parties. For example, robo-advisory providers can certainly make a contribution with webinars, podcasts, a financial encyclopedia, or a macroeconomic magazine. Even better, of course, are neutral bodies such as independent platforms or schools, taking on this task with the support of policymakers or the private sector. As people live longer, financial investment can and must play an important role in pension planning, while improvement of financial literacy can also become a key factor supporting economic growth.

13.3.3 *Greater Transparency*

Offering a complete and easy-to-understand overview of costs and related services is certainly a very important factor. However, this also includes information about the investment management approach presented in an accessible way. In addition to illustrations and images, videos, or local or virtual open-house events featuring the investment management team can play an important role.

13.3.4 *Targeting Women*

Figure 13.3 shows that 24.3% of women, but only 15.1% of men, rate the likelihood of using a robo-advisor to invest in the next 12 months as high to very high. At the same time, Figure 13.2 shows that 12.7% women vs. 28.5% men are familiar with the term robo-advisor. To an extent, this may be because the marketing campaigns and online presence of many robo-advisors is too male-focused. Individual events, information materials, and studies aimed at women can help to increase the identified potential. Parallel efforts, such as increasing the proportion of women on supervisory boards and in senior management in companies, should help women to accumulate wealth over time, which in turn is likely to play an increasingly important role in boosting the growth potential going forward.

13.3.5 Hybrid Access

In addition to the measures mentioned under the previous points for attracting previously "uninvested" participants to investment and pushing out other existing investment products, a hybrid model can play a key role in building trust. This is where robo-advisors and client advisors act in something of a tandem. For example, a potential client can first create a personal investment proposal online without obligation using the robo-advisor's investment onboarding process. The potential client can then contact a client advisor to clarify any questions regarding content or to further personalize the investment strategy. Charles Schwab and the private bank, Hauck & Aufhaeuser have implemented this model with Schwab Intelligent Portfolios PremiumTM in the US (Charles Schwab & Co. Inc. 2020) and Zeedin in Germany.

13.3.6 Expanding and Improving the Product Range

Besides the continuous improvement of the user interface (on mobile devices in particular) and the intuitiveness of the application based on customer feedback and innovative methods such as solution-based thinking, the robo-advisory industry has to ask itself what can be done to truly revolutionize the market. In view of the points mentioned above, progressive technologies such as voice-controlled applications could prove to be promising here. However, this also includes applications that introduce the user to the topic of investment depending on their particular circumstances and related interests (e.g. saving for a college fund for their children, buying a property, pensions). Playful elements that can help clients identify potential needs easily and intuitively can also be useful here.

13.4 Summary

All these measures or activities can be leveraged to raise awareness and achieve the critical mass required for robo-advisors to be profitable. As clients pay a recurring, annual fee for the robo-advisor services they use, long-term customer loyalty and the continued acquisition of new clients will be a key factor for the profitability of robo-advisors in the medium-term. Robo-advisors become profitable when the total fee received for both the volume of existing business and new business in the respective year exceeds the costs for the associated new customer acquisition (the marketing costs are typically quite considerable) and the costs for ongoing operations (e.g. personnel costs for portfolio management, customer service, middle and back office areas as well as material costs such as IT costs).

References

Antoniotti, Joe, Donna DenBraven, and Greg DePalma. 2018. *Ipsos Affluent Intelligence: Fintech: The Technology and the Early Adopters Disrupting Financial Services.* Tech. rep. Ipsos. https://www.ipsos.com/sites/default/files/ct/publication/documents/2018-05/iai-fintech-whitepaper-5-2018.pdf.

Baur, Susanne. 2018. *Einmal Ausatmen Bitte: Ikea Bringt eine Kollektion mit Adidas Raus.* Last Accessed 20 March 2019. https://www.instyle.de/fashion/ikea-adidas-kollektion.

Beardsley, Brend, Daniel Kessler, Martin Mende, Federico Muxi, Matthias Naumann, Jürgen Rogg, Tjun Tang, Tyler Woulfe, André Xavier, and Anna Zakrzewski. *Global Wealth 2018: Seizing the Analytics Advantage.* Tech. rep. P. 11. Boston Consulting Group.

Bhave, Ashwin, Chris Biggs, Peter Burggraaff, Brad Loftus, and Shishir Pathak. 2018. *Accelerating Digital Innovation in Retail.* Tech. rep. P. 7. Boston Consulting Group.

Bradtmöller, Pia, and Annabella Düchting. 2018. *J.P. Morgan Asset Management – Income-Barometer 2018.* Last Accessed 24 March 2019. http://www.jpmorganassetmanagement.de/CMSMediaStorage/RADEditorDocuments/2018-07-05_JPMAM%20Income-Barometer%20Deutschland%202018.pdf.

Brummer, Thomas. 2018. *Sebastian Hasenack (Investify): "Testphase mit Haspa beendet".* Last Accessed 20 March 2019. https://de.extraetf.com/news/finanznews/sebastian-hasenack-investify-testphase-mit-haspa-beendet.

Buess, Robert, Daniel B. Hildebrand, Dominik Löber, and Jeanette Okwu. 2018. *Rewiring Asset and Wealth Management*, P. 8. Munich: Roland Berger GMBH.

Charles Schwab & Co. Inc. 2020. *Intelligent Advisory.* Last Accessed 20 March 2019. https://www.schwab.com/public/schwab/investment_advice/intelligent_advisory.

Computer Bild. 2019. *IKEA und Sonos: Symfonisk RegalWiFi-Speaker im Test.* Last Accessed 23 March 2019. https://www.computerbild.de/artikel/cb-News-Vernetztes-Wohnen-Smarthome-IKEA-Sonos-Symfonisk-19465133.html.

Dohms, Heinz-Roger. *Immer Mehr Deutsche Banken Begraben ihre Robo-Projekte*. Last Accessed 20 June 2019. https://finanz-szene.de/digital-banking/immer-mehr-deutsche-banken-begraben-ihre-robo-projekte/.

Friedberg, Barbara. 2018. *Robo-Advisors with the Most Assets Under Management (AuM)*. Last Accessed 10 March 2019. https://www.roboadvisorpros.com/robo-advisors-with-most-aum-assets-under-management/.

Gobell, Joetta. 2019. *The Affluent Millennial Investing Survey*. Investopedia. Last Accessed 1 December 2019. https://www.investopedia.com/the-investopedia-affluent-millennials-study-4769751.

IT Finanzmagazin. 2018. *Sebastian Hasenack (Investify): "Testphase mit Haspa beendet"*. Last Accessed 20 June 2019. https://www.it-finanzmagazin.de/online-vermoegensverwalter-vaamo-und-moneyfarm-vor-zusammenschluss-81232/.

Nicolaisen, Christian. 2019. *Studie von Ebase: Nur jeder fünfte Deutsche kann etwas mit Robo-Advice anfangen*. Last Accessed 9 March 2019. https://www.private-banking-magazin.de/ebase-umfrage-nur-jeder-fuenfte-deutsche-kann-etwas-mit-robo-advice-anfangen/.

Red Bull. 2019. *Red Bull Crashed Ice Boston 2019*. Last Accessed 20 March 2019. https://www.redbull.com/se-en/events/crashed-ice-boston.

Robo-Advisor Portal. 2018. *Schluss, Aus, Vorbei: Werthstein schließt die Pforten!* Last Accessed 20 March 2020. https://www.roboadvisor-portal.com/schluss-aus-vorbei-werthstein-schliesst-die-pforten/.

Samann, Lena. 2019. *Warum Sie Jetzt ein Design von Virgil Abloh bei IKEA Kaufen Können*. Last Accessed 20 March 2020. https://www.vogue.de/mode/artikel/virgil-abloh-ikea.

Sampiero, Josh. 2013. *Flieg mit Baumgartner: Red Bull Stratos-Video*. Last Accessed 20 March 2019. https://www.redbull.com/de-de/flieg-mit-baumgartner-red-bull-stratos-pov-video.

TagesgeldVergleich. 2019. *Geldvermögen Privater Haushalte*. Last Accessed 31 March 2020. https://www.tagesgeldvergleich.net/statistiken/geldvermoegen.html#entwicklung.

Test. 2018. *Was die Automatisierte Vermögensverwaltung Taugt*. Last Accessed 20 March 2019. https://www.test.de/Robo-Advisor-Was-die-automatisierte-Vermoegensverwaltung-taugt-5107535-0/.

Verbrauchermagazin-Redaktion. 2017. *Umfrage: Die Deutschen überschätzen ihr Finanzwissen*. Last Accessed 20 March 2019. https://bankenverband.de/blog/umfrage-die-deutschen-uberschatzen-ihr-finanzwissen/.

Wenig, Mirko. 2018. *Deutsche Parken Geld am Liebsten auf Sparkonto*. Last Accessed 20 March 2019. https://www.versicherungsbote.de/id/4862540/Deutsche-parken-Geld-auf-Sparkonto/.

Glossary

Active management Whereas most robo-advisors follow a passive approach, that is, to follow indices as closely as possible, some robos also apply active management, that is, they are trying to outperform a given benchmark.

Application programming interface (API) Intermediary interface between multiple parts of software in order to facilitate ease of implementation, execution, and maintenance of interrelated programs.

Artificial intelligence Artificial Intelligence (AI) is a broadly used term to describe the capability of a computer system, which resembles cognitive functionality, especially in decision making. General AI denotes a type of artificial intelligence that is not bound to a certain task and mimics human behavior. It is, however, not yet existent. Applied AI, on the other hand, is focused on solving specific real world problems, mostly in a data-driven way and makes use of machine learning algorithms.

Asset allocation The distribution of assets in a portfolio to diversify the wealth of the investor. Typical asset classes comprise stocks, bonds, and cash. Further asset classes for robo-advisory may include real estate, commodities (mostly gold), hedge funds, and alternative investments.

Behavioral finance A theory, which describes the impact from emotional biases on financial decision-making. By and large, the findings from social psychology are adapted to similar situations in finance to derive conclusion about the behavior of investors. Financial models

© The Author(s) 2021

P. Scholz (ed.), *Robo-Advisory*, Palgrave Studies in Financial Services Technology, https://doi.org/10.1007/978-3-030-40818-3

often assume rational agents on financial markets, the concept of homo economicus.

Black-Litterman model The Black–Litterman model is a mathematical model for portfolio allocation by Fischer Black and Robert Litterman published in 1992. It takes into account uncertainties in risk and (most important) return estimators and gives therefore more robust results than the older Markowitz model (simple mean-variance optimization).

Chatbot A computer system that deals with digitalized written or spoken language and follows a certain conversation protocol, based on the provided textual or spoken content. The goal is to provide a human-machine interface to solve specific problems, such as product-related questions. Hence, for example, chatbots interact with people in sales or support scenarios, but manifold other areas of interaction are possible.

Compound annual growth rate (CAGR) The CAGR is a measure for the average growth over a multitude of periods.

Core banking system The core banking system is a software, which stores and processes the core data of a bank.

Cross-selling campaign Cross-selling denotes a process in the customer-vendor interaction, in which the customer is recommended additional items to buy, typically following a pre-defined sales strategy. For example, after purchasing a pair of shoes, material for shoe care is offered to the customer.

Customer journey The entire sequence of touch points a customer experiences when interacting with a company's offering, from awareness to purchase to advocacy.

Customer-salesperson interaction technologies Technologies that enable customers to inform about product and services completely on its own.

Diversification breakdown In "normal" times different risky assets will behave differently. This is the reason why the total risk of a portfolio which consists of different asset classes lies normally well below the weighted sum of the individual risks. During crisis, however, investors tend to get out of their risky positions globally and simultaneously and therefore this diversification effect is vanishing in times when it is needed most.

Diversification ratio The diversification ratio is defined as the ratio of the weighted average of volatilities divided by the portfolio volatility. In other words, it is the portfolio risk without taking diversification into account divided by the portfolio risk with diversification.

Exchange traded fund (ETF) An ETF, very often also called index fund, is an investment fund, which tries to minimize the tracking error compared to a given benchmark. It is hence used for passive investing and is part in many robo-advisory portfolios.

FinTech Companies which provide financial services based on modern technologies are denoted as FinTechs. Very often, they are start-ups that offer just one specific service and not a whole range of services. The idea of FinTech is rather old because there has always been a tendency to use innovations for financial purposes to gain advantage over other market participant, for example, the telegraph for transmitting stock market information.

General data protection regulation (GDPR) The GDPR is an EU law, which regulates data protection and privacy in most of the European countries.

Home bias Describes an emotional bias in decision-making. Following academic research, investors have a tendency to overweight domestic assets and to underweight foreign ones. This effect can be found with private as well as professional investors and creates portfolios with limited diversification benefits.

Hybrid model Robo-Advisors that integrate a human advisor as well in their process are called hybrid because they are link two different approaches: the strength of the robot by implementing an automatized investment process, as well as a human for personal contact and empathy if necessary, for example, during market downturns.

Irrational behavior Decisions and choices which do not maximize utility, against the classical economic theory assumption of rationality.

Mental accounting Describes the tendency to separate different investments into different internal accounts. In some cases, this can be quite useful, for example, if different investment goals are pursued. But if the economic situation is the same, sometimes still different decisions are made—which would be inconsistent. Mental accounting links to the Prospect Theory of Kahneman and Tversky.

Markets in financial instruments directive (MiFID) Established in 2007, MiFID is a European regulation to improve transparency and efficiency of the European financial markets. It also determines standards for regulatory disclosures. Meanwhile, there is an updated regulation in act, which is called MiFID 2.

Minimum viable product The minimum viable product denotes an early version of a product, which typically includes core features only.

The idea is that customers like early adopters provide enough feedback for successful enhancements of the product.

Omni-channel Denotes a sales strategy, which includes the use of several distribution channels. For the improvement of customer experience, the user decides which channel to use for a certain service and can switch between the different channels seamlessly.

Overconfidence Describes the observation that people tend to overestimate their knowledge or abilities. Overconfidence is a sub-form of control illusion. Investors who are suffering from overconfidence may estimate the investment's risk incorrectly and may create portfolios with poor diversification.

Passive management In contrast to active portfolio management, passive management tries to track a given benchmark as closely as possible. The aim is to recreate, for example, an index return as accurate as possible and to avoid any deviations. Exchange Traded Funds (ETFs) are a popular instrument for passively managed portfolios.

Payment services directive 2 (PSD 2) European directive aiming at furthering the integration of the market for electronic payments within the EU in order to increase competition and attain better prices for consumers.

Perceived risk Consumer's subjective belief of suffering a loss in the pursuit of desired outcomes.

Perceived use Individual's perception that the use of a new technology will enhance or improve his or her performance.

Psychological reactance Threat of perceived freedom by elimination.

Questionnaire Set of questions, which is usually applied to measure the risk profile of investors in order to give adequate investment advice.

Rationality In financial models, it is very often assumed that fully rational investors are decision-makers. This comprises the assumption of no emotional biases, complete information, and a stable utility function.

Risk In contrast to return, risk is the opponent in the story of investing. Risk can be measured in different ways: it can be the deviation from a given benchmark—no matter in which direction; or it can be the potential loss an investor might bear.

Risk capacity Amount of risk an investor could bear based on his or her wealth or investment requirements. For example, investors with a longer investment horizon could accept higher equity levels than short-term investors.

Risk parity This investment approach seeks to optimize the balance of risk within a portfolio by assigning each asset the same risk budget.

Risk profile In the risk profile, the risk tolerance and risk capacity of an investor is measured, typically by a questionnaire.

Risk tolerance Amount of risk an investor could bear based on his or her risk appetite: from risk-averse to risk-loving, depending on individual preferences. However, people who love to take risks do not necessarily prefer risk in every field.

Robo-advice Digital investment advice tools that match customers on the base of their personal preference to financial products.

Robo-advisor A financial service based on two components: investment advice and automatization through robots. The term robo-advice may comprise different levels of automatization: whereas the onboarding and risk profiling is typically provided by algorithms, the investment process can either be completely rule-based or driven by committees.

SEC The US Securities and Exchange Commission (SEC) is a federal agency. It is responsible for the regulation of markets and exchanges.

Self-profiling A process through which financial advisory clients create a risk-tolerance, risk-budget profile of themselves, aided by automated computer systems.

Social media Describes platforms, which allow internet users to communicate and hence to share user-generated content like knowledge, opinions, evaluations, impressions and so on.

Trust Confidence in the exchange partner's reliability and integrity.

White-label Financial service or product, which is offered without the brand of the originating bank.

Index

© The Author(s) 2021

P. Scholz (ed.), *Robo-Advisory*, Palgrave Studies in Financial Services Technology, https://doi.org/10.1007/978-3-030-40818-3

The manufacturer's authorised representative in the EU is Springer Nature Customer Service Centre GmbH, Europaplatz 3, 69115 Heidelberg, Germany. If you have any concerns regarding our products, please contact ProductSafety@springernature.com

Printed and bound by CPI Group (UK) Ltd, Croydon, CR0 4YY
30/09/2026
02225461-0011